Springer Theses

Recognizing Outstanding Ph.D. Research

For further volumes:
http://www.springer.com/series/8790

Aims and Scope

The series "Springer Theses" brings together a selection of the very best Ph.D. theses from around the world and across the physical sciences. Nominated and endorsed by two recognized specialists, each published volume has been selected for its scientific excellence and the high impact of its contents for the pertinent field of research. For greater accessibility to non-specialists, the published versions include an extended introduction, as well as a foreword by the student's supervisor explaining the special relevance of the work for the field. As a whole, the series will provide a valuable resource both for newcomers to the research fields described, and for other scientists seeking detailed background information on special questions. Finally, it provides an accredited documentation of the valuable contributions made by todays younger generation of scientists.

Theses are accepted into the series by invited nomination only and must fulfill all of the following criteria

- They must be written in good English.
- The topic should fall within the confines of Chemistry, Physics, Earth Sciences, Engineering and related interdisciplinary fields such as Materials, Nanoscience, Chemical Engineering, Complex Systems and Biophysics.
- The work reported in the thesis must represent a significant scientific advance.
- If the thesis includes previously published material, permission to reproduce this must be gained from the respective copyright holder.
- They must have been examined and passed during the 12 months prior to nomination.
- Each thesis should include a foreword by the supervisor outlining the significance of its content.
- The theses should have a clearly defined structure including an introduction accessible to scientists not expert in that particular field.

Kristiaan De Greve

Towards Solid-State Quantum Repeaters

Ultrafast, Coherent Optical Control
and Spin-Photon Entanglement
in Charged InAs Quantum Dots

Doctoral Thesis accepted by Stanford University, USA

 Springer

Kristiaan De Greve
Department of Physics
Harvard University
Cambridge, MA
USA

Supervisor
Yoshihisa Yamamoto
Edward L. Ginzton Laboratory
Stanford University
Stanford, CA
USA

ISSN 2190-5053 ISSN 2190-5061 (electronic)
ISBN 978-3-319-00073-2 ISBN 978-3-319-00074-9 (eBook)
DOI 10.1007/978-3-319-00074-9
Springer New York Heidelberg Dordrecht London

Library of Congress Control Number: 2013934550

© Springer International Publishing Switzerland 2013

This work is subject to copyright. All rights are reserved by the Publisher, whether the whole or part of the material is concerned, specifically the rights of translation, reprinting, reuse of illustrations, recitation, broadcasting, reproduction on microfilms or in any other physical way, and transmission or information storage and retrieval, electronic adaptation, computer software, or by similar or dissimilar methodology now known or hereafter developed. Exempted from this legal reservation are brief excerpts in connection with reviews or scholarly analysis or material supplied specifically for the purpose of being entered and executed on a computer system, for exclusive use by the purchaser of the work. Duplication of this publication or parts thereof is permitted only under the provisions of the Copyright Law of the Publisher's location, in its current version, and permission for use must always be obtained from Springer. Permissions for use may be obtained through RightsLink at the Copyright Clearance Center. Violations are liable to prosecution under the respective Copyright Law.

The use of general descriptive names, registered names, trademarks, service marks, etc. in this publication does not imply, even in the absence of a specific statement, that such names are exempt from the relevant protective laws and regulations and therefore free for general use.

While the advice and information in this book are believed to be true and accurate at the date of publication, neither the authors nor the editors nor the publisher can accept any legal responsibility for any errors or omissions that may be made. The publisher makes no warranty, express or implied, with respect to the material contained herein.

Printed on acid-free paper

Springer is part of Springer Science+Business Media (www.springer.com)

Supervisor's Foreword

At the time of writing of this dissertation, the future of quantum information processing research, and in particular that of currently proposed quantum computing machines, is still elusive. The following is the summary of the current majority opinions in the scientific community (end of 2012). Any physical qubit has still a too short decoherence time compared to expected/required computational times for meaningful tasks, such as factoring of 1,024-bit integer numbers or quantum entanglement distribution over 1,000 km distance. Any current physical gate operation is faulty, and leads to computational errors, that need to be accounted for. The only existing solution for circumventing these two problems is the use of quantum error correcting codes, and fault-tolerant quantum computing architectures.

A recent theoretical study on a layered quantum computing architecture with a topological surface code (N.C. Jones *et al.*, *Physical Review X*, 2, 031007 (2012)) uncovers the prospective system size of such fault-tolerant quantum computers. The required gate fidelity still exceeds 99.9 %, and the number of physical qubits is 10^8–10^9, with an overall computational time as long as 1–10 days for factoring a relatively small (1,024-bit) integer number, or for quantum simulating a relatively small molecule with only 60 electrons and nuclei.

How to physically implement such a huge quantum computer with numerous qubits? One is tempted to propose a distributed quantum information processing system connected by entangled memory qubits and quantum teleportation protocols. However, if we evaluate the resources required for high-fidelity entanglement distribution over non-local memory qubits, we can easily convince ourselves that a distributed quantum information processing network is not a practical solution. The overall computational time would be many years for factoring a 1,024-bit integer number instead of around 1 day. We must integrate 10^8–10^9 physical qubits into one chip in order to avoid this serious communication bottleneck and construct a useful quantum computer.

Advanced molecular beam epitaxy and nanolithography techniques for optical semiconductors now allow us to grow InAs quantum dots (QDs) in GaAs host matrices or even in GaAs/AlAs microcavities in a square lattice geometry with regular spacing of 100–1,000 nm (C. Schneider et al., *Applied Physics Letters* 92, 183101 (2008)). This means that 10^8–10^9 QDs can be readily integrated into a reasonable 1 cm^2 chip. Such an optically active semiconductor QD can trap a single electron or hole as a matter (spin) qubit (M. Bayer et al., *Physical Review B* 65, 041308 (2002)), and simultaneously emit a single photon as a communication qubit (P. Michler et al., *Science* 290, 2282 (2000)).

This particular system of an InAs QD embedded in a GaAs/AlAs microcavity is the platform on which Kristiaan De Greve has conducted various experiments in my research group while working toward his PhD thesis at Stanford University. Before Kristiaan started his PhD thesis work in my group, we had accumulated some knowledge and techniques in this field. A Fourier-transform-limited single photon wavepacket, which is a quantum mechanically indistinguishable particle and an indispensible resource for quantum teleportation and quantum repeater systems, was generated from a single InAs QD in a micropost-microcavity (C. Santori et al., *Nature* 419, 594 (2002)). An entangled photon-pair can be produced by the collision of these two sequentially generated single photons at a 50–50 beam splitter, for which we demonstrated the violation of a Bell's inequality. Indistinguishable single photons can also be generated by two independent emitters using another optically active compound semiconductor, ZnSe.

We had managed to manipulate a single electron spin in an InAs QD by off-resonant stimulated Raman scattering using single picosecond optical pulses, by which a general SU(2) operation for an electron spin can be implemented within tens of picoseconds (D. Press et al., *Nature* 456, 218 (2008)). Using Ramsey-interometry, the dephasing time (T_2^*) of a donor bound electron had also been measured to be a few ns. By virtue of a Hahn-spin-echo protocol, this noise source could be decoupled, resulting in a decoherence time (T_2) of a few microcseconds. This is where Kristiaan's research adventure started: with a project to implement an optical refocusing pulse technique to increase the decoherence time of a single quantum dot electron spin (D. Press, K. De Greve et al., *Nature Photonics* 4, 367 (2010)). He then moved on to second project, in line with the former one, to demonstrate a quantum dot hole spin qubit which enjoys a suppressed hyperfine interaction with In and As nuclear spins (K. De Greve et al., *Nature Physics* 7, 872 (2011)), to end with a third major project: a system-level experiment to generate and demonstrate an entangled state of a single photon and a single spin (K. De Greve et al., *Nature* 491, 421 (2012)).

Stanford, CA, USA   Yoshihisa Yamamoto

Summary of the Dissertation

Single spins in optically active semiconductor host materials have emerged as leading candidates for quantum information processing (QIP). The quantum nature of the spin degree of freedom allows for encoding of stationary, memory quantum bits (qubits), and their relatively weak interaction with the host material preserves the coherence between the spin states that is at the very heart of QIP. On the other hand, the optically active host material permits direct interfacing with light, which can be used both for all-optical manipulation of the quantum bits, and for efficiently mapping the matter qubits into flying, photonic qubits that are suited for long-distance communication. In particular, and over the past two decades or so, advances in materials science and processing technology have brought self-assembled, GaAs-embedded InAs quantum dots to the forefront, in view of their strong light-matter interaction, and good isolation from the environment. In addition, advanced and established microfabrication techniques allow for enhancing the light-matter interaction in photonic microstructures, and for scaling up to large-size systems.

One of the (as of yet) most successful applications of QIP resides in the distribution of cryptographic keys, for use in one-time-pad cryptographic systems. Here, the bizarre laws of quantum mechanics allow for clever schemes, where it is in principle impossible to copy or obtain the key (as opposed to practically, computationally hard schemes used in current, 'classical' schemes). Proof-of-principle schemes were demonstrated using transmission of single photons, though unavoidable photon losses and limited efficiency of the detectors used limit their use to distances of several hundred kilometers at most. Longer-range systems will need to rely on massively parallel, pre-established links consisting of quantum mechanically entangled memory qubits, with the information transfer occurring through quantum teleportation: the so-called quantum repeater. The establishment of such entangled qubit pairs relies on the possibility to efficiently map quantum information from memory qubits to flying, photonic qubits – the realm of charged, InAs quantum dots.

This work elaborates on previously established all-optical coherent control techniques of individual InAs quantum dot electron spins, and demonstrates proof-of-principle experiments that should allow the utilization of such quantum dots for future, large-scale quantum repeaters. First, we show how more elaborate, multi-pulse spin control sequences can markedly increase the fidelity of the individual spin control operations, thereby allowing many more such operations to be concatenated before decoherence destroys the quantum memory. Furthermore, we implemented an ultrafast, gated version of a different type of control operation, the so-called geometric phase gate, which is at the basis of many proposals for scalable, multi-qubit gate operations. Next, we realized a new type of quantum memory, based on the optical control of a single hole (pseudo-)spin, that was shown to overcome some of the detrimental effects of nuclear spin hyperfine interactions, which are assumed to be the predominant sources of decoherence in electron spin-based quantum memories – at the expense, however, of a larger sensitivity to electric field-related noise sources.

Finally, we discuss a system-level experiment where the quantum dot electron spin is shown to be entangled with the polarization of a spontaneously emitted photon after ultrafast, time-resolved (few picoseconds) downconversion to a wavelength (1,560 nm) that is compatible with low-loss optical fiber technology. The results of this experiment are two-fold: on the one hand, the spin-photon entanglement provides the necessary light-matter interface for entangling remote memory qubits; on the other hand, the transfer to a low-fiber-loss wavelength enables a significant increase in the potential distance range over which such remote entanglement could be established. Together, these two aspects can be seen as a necessary preamble for a future quantum repeater system.

Acknowledgements

This dissertation is the result of several years of research conducted at Stanford, where I had the honor to meet and work with some of the most talented people one can imagine – people who helped and inspired me, encouraged and corrected me when needed (often, in the latter case), and provided the proverbial 'shoulders of giants' on which it is a pleasure to stand. First and foremost, I should thank my advisor, Yoshihisa Yamamoto, for the incredibly open and stimulating environment that I and other students in his group have been enjoying. Yoshi's approach is one in which students are encouraged and given the freedom to study problems very much in depth, all the while making sure not to forget about the big picture. It is his ability and emphasis to discern truly important problems from the low-hanging fruit that has probably impressed me the most while I was peripatetically wandering around in his group, seeking out interesting problems to solve. I would also like to thank the other members of my reading committee, Jelena Vuckovic and Mark Brongersma, who are both excellent teachers and research mentors in their own right. I very much enjoyed interacting with them and their research groups, and their presence at Stanford was an important factor in my decision to tackle graduate studies here. Hideo Mabuchi and Mark Kasevich, with their deep insights in quantum information, cavity-QED and atomic physics, were truly inspiring teachers, and I really appreciated their willingness to serve on my defense committee.

Within the Yamamoto group, Thaddeus Ladd, David Press and Peter McMahon have probably been my closest day-to-day collaborators. Thaddeus combines an incredible insight in all things quantum, with a wide-ranging and open-minded curiosity that makes it a pleasure for anyone to work with and be mentored by him. Dave Press is one of the finest physicists and experimenters that I have ever met, and him taking me under his wings and allowing me to collaborate on his final projects was very important for me. Most of the experimental techniques used in this dissertation were developed or fine-tuned by Dave, and his attention for details and emphasis on doing challenging experiments in the cleanest, best way possible is something I very much admire and hope to emulate. Peter also combines fine experimental skills with a sharp and critical mind – a combination that makes him a highly valued collaborator and labmate, and one that will enable him to pursue a variety of challenging problems in the future.

During the final part of my PhD research, I collaborated quite intensively with Jason Pelc, Leo Yu, Chandra Natarajan and Marty Fejer. Jason and Marty are the wizards of non-linear optics, and together with Leo and Chandra, they were instrumental for the realization of the time-resolved downconverters that formed the cornerstone of the spin-photon entanglement experiments. Leo and Chandra will continue this line of research in the near future, which should allow to overcome limitations to upscaling of the quantum dot system due to dot-to-dot inhomogeneities. Nathan Cody Jones always offers fresh and original views on quantum information, and his theoretical enthusiasm forms a nice counterbalance for the critical realism of the cynical experimenters – I hope that his view may prevail, and that one day we will indeed see a quantum computer at work. Na Young Kim and Darin Sleiter contributed significantly to this work, each in their own, behind-the-scenes, selfless way. I am also grateful to all the other members and former members of the Yamamoto group, with whom it has been a pleasure to interact: Glenn Solomon, Shinichi Koseki, Kai-Mei Fu, Susan Clark, Qiang Zhang, Zhe Wang, Kaoru Sanaka, Wolfgang Nitsche, Eisuke Abe, Benedikt Frieß, Kai Wen, George Roumpos, Michael Lohse, Jung-Jung Su, Shruti Puri, Katsuya Nozawa, Tomoyuki Horikiri, Bingyang Zhang, Patrik Recher, Lin Tian, Chih-Wei Lai, Stephan Götzinger, Hiroki Takesue, Mike Fraser, Tim Byrnes, Parin Dalal, Hui Deng, Eleni Diamanti, Neil Na, Sheelan Tawfeeq, Crystal Bray, Cyrus Master, as well as all our colleagues from the National Institute of Informatics and at Nihon University. Among the latter, Naoto Namekata and Shuichiro Inoue provided us with much appreciated ultra-low-noise single-photon telecom-wavelength detectors.

Sven Höfling and his colleagues in the Forchel group in Würzburg (Christian Schneider, Dirk Bisping, Sebastian Maier and Martin Kamp to mention only a few of them) provided the excellent quantum dot samples without which none of this research would have been possible. I particularly appreciated the stimulating discussions with Sven during his numerous visits to California, which went far beyond quantum dot growth per se.

Throughout my PhD research, it was a pleasure to be able to discuss with many current and former members of the Vuckovic group, who share a common interest in all things scientific, and in particular, the future of quantum information science: Andrei Faraon, Dirk Englund, Ilya Fushman, Yiyang Gong, Brian Ellis, Arka Majumdar, Erik Kim, Michal Bajcsy, Konstantinos Lagoudakis and Tom Babinec among others.

Takahiro Inagaki and Hideo Kosaka from Tohoku University visited our lab last year, and contributed significantly to the geometric phase-gate experiments.

While not reported in this dissertation, several people contributed to the various side-projects which I very much enjoyed tackling. For the ZnSe experiments at the very beginning of my joining the Yamamoto group, Alex Pawlis from the university of Paderborn was the driving force, while Ian Fisher and Jiun-Haw Chu offered me the opportunity to learn a lot about (and contribute a very small amount to) the study of a new class of high-T_C superconductors.

Acknowledgements

Among the many excellent members of the technical and administrative staff at Stanford, Yurika Peterman and Rieko Sasaki stand out in view of their tireless dedication and kind attention to detail. I am also indebted to the Ginzton front office staff and the EE department's administrative staff.

In some sense, a PhD is the culmination of many years of study. I had the privilege of learning from and being mentored by excellent people back home, at KU Leuven. Some of the people I would like to especially acknowledge in this regard are profs. Jo De Boeck, Robert Mertens, Karen Maex, Staf Borghs and Hugo De Man, and Drs. Wim Van Roy, Liesbet Lagae and Pol Van Dorpe. The financial support of the Belgian American Educational Foundation, and from the Stanford Graduate Fellowship program (Dr. Herb and Jane Dwight fellowship) offered me the financial independence that, directly and indirectly, enabled much of the research in this dissertation.

Throughout my time at Stanford, I enjoyed the company of good friends, and listing all of them would be quite daunting. Nevertheless, I would like to especially mention Thomas Tsai, Jim Loudin, Sabina Alistar, Jessica Faruque, Adrian Albert, Gaurav Bahl, Daniel Barros, Rita Lopez, Dany-Sébastien Ly-Gagnon, Clara Kuo, Punya Biswal, Smita Gopinath, Shrestha Basu Mallick, Viksit Gaur, Rinki Kapoor, Benjamin Armbruster, Iwijn De Vlaminck, Katja Nowack, Sophie Walewijk, Lieven Verslegers and Tracy Fung, who made life on the Farm a very pleasant experience. In addition, many old friends from Leuven and Schilde made coming home during the holidays a very pleasant experience: Walter Jacob, Reinier Vanheertum, Marlies Sterckx, Pol Van Dorpe, Julita Jarmuz, Johan Reynaert, Brik Peeters, Loes Lysens, Filip Logist, Katleen Hoorelbeke, Geert Gins, Anja Vananroye, Bart Creemers, Jiqiu Cheng, Geert Vermeulen and many others.

My parents are the ones who ultimately allowed my sister, my brother and me to enjoy the benefits of a much appreciated education. I will be forever indebted to my mother and father for allowing me to pursue my dreams, and hope to never have to disappoint them. My brother, sister and brother-in-law, have each been supportive in their own way. My dear grandmother passed away just before defending my dissertation – I would like to dedicate this work to her. But perhaps most importantly, my tenure at Stanford allowed me to find a soulmate, Serena. To all of you: thank you.

Contents

Foreword xiv

Summary of the Dissertation xiv

Acknowledgements xiv

1 Introduction: Solid-State Quantum Repeaters 1
2 Quantum Memories: Quantum Dot Spin Qubits 25
3 Ultrafast Coherent Control of Individual Electron Spin Qubits 39
4 All-Optical Hadamard Gate: Direct Implementation of a Quantum Information Primitive 67
5 Fast, Pulsed, All-Optical Geometric Phases Gates 75
6 Ultrafast Optical Control of Hole Spin Qubits: Suppressed Nuclear Feedback Effects 83
7 Entanglement Between a Single Quantum Dot Spin and a Single Photon 99
8 Conclusion and Outlook 119

A Fidelity Analysis of Coherent Control Operations 125

B Electron Spin-Nuclear Feedback: Numerical Modelling 129

C Extraction of Heavy-Light Hole Mixing Through Photoluminescence ... 137

D Numerical Modeling of Ultrafast Coherent Hole Rotations 139

E Hole Spin Device Design.. 143

F Ultrafast Quantum Eraser: Expected Visibility/Fidelity 147

List of Figures

Fig. 1.1	General outline of the key distribution problem in cryptography	2
Fig. 1.2	Bloch-sphere representation of a qubit/pseudospin	4
Fig. 1.3	Basic outline of an entanglement-swapping procedure	9
Fig. 1.4	Basic outline of a quantum-teleportation procedure	10
Fig. 1.5	Schematic outline of a beamsplitter	12
Fig. 1.6	Overview of the BB84 QKD protocol	14
Fig. 1.7	Schematic of the BBM92 protocol.	16
Fig. 1.8	Schematic of the first ionic teleportation experiment.	18
Fig. 1.9	Operation principle of a quantum repeater.	19
Fig. 1.10	Basic ingredients for a quantum repeater.	20
Fig. 2.1	Self-assembled quantum dots	26
Fig. 2.2	Excitation and recombination processes in self-assembled quantum dots.	28
Fig. 2.3	Outline of RF spin control	30
Fig. 2.4	Level structure of singly charged quantum dots.	32
Fig. 2.5	Coherent manipulation of a Λ-system	34
Fig. 2.6	Ultrafast stimulated Raman transitions	36
Fig. 3.1	Generalized 3-level structure	40
Fig. 3.2	Full four-level structure of an electron-charged quantum dot in Voigt geometry	42
Fig. 3.3	Coherent control as AC-stark shift	44
Fig. 3.4	Optical pumping for quantum bit initialization and readout	46
Fig. 3.5	All-optical electron spin qubit control	48
Fig. 3.6	Device design for all-optical control of a single electron spin	49
Fig. 3.7	Laboratory setup used for all-optical spin control	50
Fig. 3.8	Rabi-oscillations of a single electron spin qubit	51
Fig. 3.9	Ramsey fringes of a single electron spin qubit	53
Fig. 3.10	Full control over the surface of the Bloch sphere	54
Fig. 3.11	Overlap of the electron wavefunction and the nuclear spins	55

Fig. 3.12	Runners analogy of ensemble dephasing and T_2^*-effects	56
Fig. 3.13	Effect of dynamic nuclear polarization on electron-spin Ramsey fringes	57
Fig. 3.14	Feedback mechanism giving rise to dynamic nuclear polarization under all-optical electron spin control	58
Fig. 3.15	Numerical modelling of the non-linear feedback between a single electron spin and an ensemble of nuclear spins	60
Fig. 3.16	Runners analogy of spin echo	62
Fig. 3.17	All-optical spin echo for a single electron spin	63
Fig. 3.18	T_2^*-decay, visualized by means of a spin echo	64
Fig. 4.1	Imperfect Rabi-oscillations due to off-axis rotation pulses	68
Fig. 4.2	Finite-duration rotation pulses resulting in off-axis spin rotations	69
Fig. 4.3	Hadamard pulses and composite π-pulses	70
Fig. 4.4	Schematic of a 4f-grating shaper, used for pulse stretching	71
Fig. 4.5	Ramsey interferometry with Hadamard gates	72
Fig. 4.6	Composite, Hadamard-based π-pulses for spin echo	73
Fig. 5.1	Global phase for a cyclic, 2-level transition	76
Fig. 5.2	Visualing the global phase in a Ramsey interferometer	78
Fig. 5.3	The geometric phase in a Ramsey interferometer	79
Fig. 5.4	The net geometric phase in a 4-level system	79
Fig. 6.1	Wavefunctions of electrons and hole	84
Fig. 6.2	All-optical control of a single quantum dot hole spin	86
Fig. 6.3	Device design for deterministic hole charging	87
Fig. 6.4	Deterministic hole-charging of a single quantum dot	88
Fig. 6.5	All-optical control of a single hole qubit	89
Fig. 6.6	Rabi-oscillations of a single hole qubit	89
Fig. 6.7	Ramsey-fringes of a single hole qubit	90
Fig. 6.8	Complete SU(2) control of a single hole qubit	90
Fig. 6.9	Re-emergence of hysteresis-free dynamics for hole spins	91
Fig. 6.10	T_2^* and electrical dephasing	93
Fig. 6.11	Spin echo and T_2 decoherence for a single hole qubit	94
Fig. 7.1	Spin-photon entanglement from Λ-system decay	100
Fig. 7.2	Ultrafast downconversion for quantum erasure	104
Fig. 7.3	Time-resolved downconversion: performance	105
Fig. 7.4	Spin-photon entanglement verification: overview	106
Fig. 7.5	Full system diagram of the optical setup used for spin-photon entanglement verification	107
Fig. 7.6	Spin-photon correlation histograms	108
Fig. 7.7	Spin-photon correlations in the linear basis of spin and polarization	110
Fig. 7.8	Spin-photon correlations in the rotated basis of spin and polarization	110

Fig. 7.9	Dual-rail implementation of 1,560 nm spin-photon entanglement.	113
Fig. 7.10	Realization of a 1,560 nm, polarization-entangled photonic qubit.	115
Fig. 8.1	Basic ingredients for a quantum repeater.	120
Fig. A.1	Coherent control axis and angle conventions.	126
Fig. B.1	Electron spin hysteresis in Ramsey interferometry	130
Fig. B.2	Electron spin hysteresis upon resonant absorption scanning.	132
Fig. B.3	Comparison of electron- and hole-spin Overhauser shifts.	133
Fig. C.1	Angular PL dependence, used for HH-LH mixing analysis	138
Fig. D.1	Coherent rotation modelling for hole qubits	140
Fig. E.1	Detailed layer structure of the hole devices used	144
Fig. E.2	Charge-tuneable hole devices: band line-up	145
Fig. F.1	Ultrafast downconversion: timing resolution	148

Chapter 1
Introduction: Solid-State Quantum Repeaters

Quantum Information Processing [1] (QIP), roughly defined as that branch of physics, engineering and computer science that attempts to incorporate fundamental concepts from quantum mechanics in order to augment and improve on existing information processing capabilities,[1] was initially proposed as an answer to a fundamental question in both theoretical physics and quantum chemistry: how to keep track of the gigantic state space that is present in large-scale quantum mechanical systems [2]? Such *quantum simulation* has since become the subject of an entire subfield of study [3], building on the intrinsic state space provided by quantum systems to understand fundamental properties of nature, especially in solid-state and many-body physics – properties and studies that would be intractable using classical mathematical tools based on digital computing power.

Similarly, and very much in concert with quantum simulation, another branch of QIP known as *quantum computation* [4] emerged, based on ingenious proposals that build on the full power of the Hilbert space in large-scale quantum systems to dramatically speed up the solution and/or verification of particular, 'hard' mathematical problems. The quantum enabled, exponential speedup in prime-factoring as demonstrated by Peter Shor in 1994 [5] led to a true explosion of interest in this subfield, as such prime factoring (more specifically: the assumed difficulty thereof) lies at the heart of widely used public-key cryptography systems such as the well-known RSA encryption.[2] Similarly, quadratic speedups in searches through unsorted databases were demonstrated by Lov Grover in 1996 [6].

---

[1] To quote from [1]: "the study of the information processing tasks that can be accomplished using quantum mechanical systems"

[2] An important caveat: *not* all cryptographic systems rely on the difficulty of prime number factoring. Contrary to popular belief, quantum computing systems are not 'quantum mechanical equivalents of classical computers', and in fact, their application scope is, at the time of writing this dissertation, quite limited. It is quite possible, and even likely, that a quantum mechanics based prime factoring machine would make itself instantly obsolete, when publicly announced: the obvious countermeasure in such a cryptographic arms race would be the abandonment of public-key cryptography...

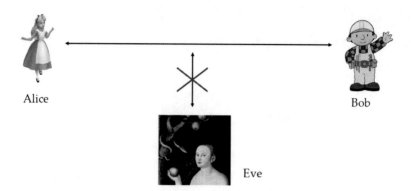

Fig. 1.1 The outline of the canonical cryptography problem: how can Alice and Bob share a secret message (or a secret key to be used in a one-time pad) without Eve being able to intercept this message?

While fascinating and enormously rich in both physics and fundamental information theory, this dissertation will for the most part steer far away from quantum computation and simulation.

Rather, we will mainly describe the submitted work within the framework of yet another branch of QIP, one that initially developed quite independently from the aforementioned ones [7, 8]: quantum communication and quantum key distribution.

The canonical problem in quantum key distribution (QKD), and quantum communication in general, is depicted in Fig. 1.1, and can be summarized as follows: how can two parties, A (Alice) and B (Bob), share secrets that cannot be overheard by an eavesdropping third party (Eve)? This problem is in some sense the opposite of the one targeted by the quantum computers trying to implement Shor's algorithm: there, quantum mechanics is used to target and break classical encryption systems, while quantum communication aims to use quantum mechanics to secure cryptographic systems [9–11].

The fact that quantum mechanics could assist in securing shared secrets may seem strange to cryptography specialists. Shortly after the second world war, Claude Shannon rigorously proved [12] the heuristics of over 50 years of cryptography and (attempted) code-breaking, in showing that a properly used one-time pad encryption system (where a truly random, truly secret cryptographic key, at least as long as the message-to-be-sent, which is used only once, is added to the message through modular addition) would be impossible to crack. Hence, when two counterparties, A(lice) and B(ob) share a mutual, secret key that satisfies the one-time pad conditions of true randomness, sufficient length and no repetition, they then can encrypt any sufficiently short message in a way that, in theory, is absolutely secure.

This moves the task of secure communication to one of sharing the secret key between the counterparties, which is where quantum mechanics can assist. It is of course possible to have, say, two disk drives with secret keys shared between Alice and Bob, from which they draw random keys as needed for communication. However, distributing these disk drives then becomes either cumbersome (if Alice and Bob are far apart: physical contact between them would require personal travel) or unsafe, as the keys would have to be sent over communication channels that are potentially unsafe. In the remainder of this chapter, we will show how several intrinsically quantum mechanical effects can assist in distributing secure, unbreakable and impossible-to-copy cryptographic keys – the realm of *quantum key distribution* [7, 8]. In the final part, we shall indicate how this thesiswork fits within the framework of a solid-state quantum repeater [13, 14], and which particular hurdles on the way to such repeaters have been overcome.

1.1 On Quantum Bits, Their Measurement, and the Inability to Clone Them

1.1.1 SU(2) and Pseudo-spins

The basic unit of QIP is the quantum bit, which is a formal, mathematical object (a state vector in a two-dimensional Hilbert space, obeying SU(2)-symmetry) that can be physically represented by a 2-level quantum system. Many such 2-level quantum systems exist and have been studied, ranging from photonic polarization states to the quantum state of a superconducting circuit, but the (arguably) canonical example of a 2-level system is a single spin-1/2 – e.g., an electron spin.[3] In a representation where our logical 0,1 become quantum states $|0\rangle, |1\rangle$ in bra-ket notation, we can map these quantum states into the up-down states of a pseudospin [15], and write the quantum state of a single qubit as follows:

$$|\Psi\rangle = \cos(\theta/2)|\downarrow\rangle + e^{i\phi}\sin(\theta/2)|\uparrow\rangle \tag{1.1}$$

Crucially, in Eq. 1.1, one can notice the existence of superpositions between the respective $|\downarrow\rangle$ and $|\uparrow\rangle$-states: the existence of *coherence* that uniquely characterizes quantum mechanics. The (pseudo-)spin-1/2 representation also allows for an intuitive representation of the quantum bit in the well-known Bloch-sphere (Fig. 1.2), where angles θ and ϕ, which define the coherence between the spin states, represent the polar and azimuthal angle with regards to the axis connecting

---

[3]In general, it can be shown that any quantum 2-level system can be mapped into a spin-1/2 representation, hence the often used *pseudo*-spin terminology [15, 16].

Fig. 1.2 The Bloch sphere representation of a 2-level quantum system, encoded as a pseudo-spin qubit

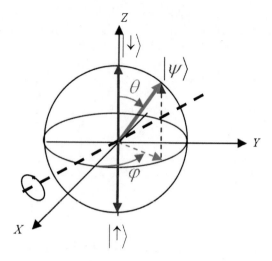

the North- and South pole – the latter corresponding to the $|\downarrow\rangle$ and $|\uparrow\rangle$-states respectively.

In view of the SU(2) symmetry, any coherent manipulation of a qubit can be described as a Hilbert-space operator in terms of the well-known Pauli spin matrices $\sigma_{x,y,z}$:

$$R_{\vec{n}}(\theta) = e^{-i\theta(\vec{n}\cdot\vec{\sigma})/2}, \sigma_x = \begin{pmatrix} 0 & 1 \\ 1 & 0 \end{pmatrix}, \sigma_y = \begin{pmatrix} 0 & -i \\ i & 0 \end{pmatrix}, \sigma_z = \begin{pmatrix} 1 & 0 \\ 0 & -1 \end{pmatrix} \quad (1.2)$$

As the notation in Eq. 1.2 suggests, coherent qubit manipulation (modulo overall complex phase factors) can be described as rotations around a rotation axis $\vec{n}$ in the Bloch-sphere (dashed line in Fig. 1.2), with rotation angle θ. Hence, in the remainder of this work, we shall adopt the terminology of coherent *rotations*.

While this semi-classical rotation picture is quite powerful and allows for an intuitive approach to qubit control, some differences do exist with classical rotations, and a particular example manifests itself in the presence of global, geometric phases upon coherent rotation. More specifically, for e.g. a 2π rotation around an arbitrary axis in the Bloch sphere, straightforward application of the rotation operator in Eq. 1.2 leads to an over-all phase factor of -1. This is a general consequence of the SU(2) symmetry and the spin-statistics theorem, and can be visualized in particular interference experiments, such as those described in Chap. 5.

Coherent manipulation and evolution is a powerful concept, at the very heart of quantum mechanics and quantum information science, yet coherences are also fragile due to the *collapse* of the wavefunction upon measurement [17, 18]. Measurement, in contrast to coherent evolution, is a non-unitary, non-reversible

1.1 On Quantum Bits, Their Measurement, and the Inability to Clone Them

and non-deterministic process; at a simplistic level, it can be described in terms of projective, Hermitian measurement operators, with only a discrete set of possible outcomes (corresponding to the eigenvalues of the operator). In the context of a Bloch-sphere representation of qubits, the measurement process can be visualized as a projection process, with the measurement operator as a particular axis in the Bloch sphere (say, X, or Y axis), and the result of the measurement being a new state vector either along $+X$ ($+Y$), or $-X$ ($-Y$), with eigenvalues ± 1. Crucially, this process is probabilistic rather than deterministic, with the probability of obtaining a particular result proportional to the angle between the qubit and the measurement axis.[4]

Quantum mechanical measurements correspond to physical observables that can, in principle, be measured experimentally; hence the requirement for Hermiticity of the measurement operators, ensuring real eigenvalues (measurement results). For a (pseudo-)spin, the most natural measurement is the one referring to the orientation of the spin; the measurement operators are again the Pauli-spin matrices. Such a measurement can be performed both directly (Stern-Gerlach-like experiment, with the magnetic field oriented in any arbitrary direction to measure the spin along any arbitrary axis), or indirectly – in the latter case, a fixed-axis spin measurement is preceded by a coherent operation that rotates the spin around another axis. The latter combination of incoherent measurement preceded by coherent rotation can be seen as an effective change of measurement basis. A concrete example: for a spin measurement in the Y-basis of the Bloch-sphere ($|\downarrow\rangle_Y, |\uparrow\rangle_Y$), a coherent rotation of $\theta = \pi/2$ around the X-axis realizes the effective measurement basis change, followed by a Z-basis measurement.

It is important to note that, in general, measurements do not preserve coherent superpositions. This can be easily seen through Eq. 1.1 and Fig. 1.2: for measurement of our qubit $|\Psi\rangle$ along the Z-axis, with $|\downarrow\rangle, |\uparrow\rangle$ as the eigenstates of the measurement operator, the resulting state vector $|\Psi_{meas}\rangle$ is either $|\downarrow\rangle$, with probability $\cos^2(\theta/2)$, or $|\uparrow\rangle$, with probability $\sin^2(\theta/2)$; all coherence is lost upon measurement. The same goes for measurements along e.g. the Y-axis ($|\downarrow\rangle_Y, |\uparrow\rangle_Y$ as eigenstates), where coherence between the $|\downarrow\rangle_Y, |\uparrow\rangle_Y$-states would be lost. Interestingly, the resulting state vector (say, $|\uparrow\rangle_Y$), while an eigenvector of the spin-Y-measurement operator, is now a superposition of Z-eigenstates: with Eq. 1.1, $|\uparrow\rangle_Y = \frac{1}{\sqrt{2}}(|\downarrow\rangle_Z + i|\uparrow\rangle_Z)$. In other words: what is a coherence in one basis, becomes an eigenstate in another – again illustrating the close relationship between coherent rotations and measurement basis change.[5]

---

[4]This description sidesteps the deep yet also deeply philosophical question about the reality of the state vector/wavefunction. In the present work, we adhere to a heuristic, Copenhagen-like interpretation of the wavefunction, generally summarized as "shut up and calculate". Note also that our definition of measurement is, strictly speaking, only valid for a so-called *strong*, projective measurement, and not for weak, partial measurements.

[5]Strictly speaking, this is only true for *pure* states, that can be represented as unit-length state vectors within the Bloch sphere; for *mixed* states, no coherent rotation or basis change can result in an eigenstate. In general, a density-matrix description [18] is required to properly deal with non-pure states.

In view of the Heisenberg uncertainty principle [17], subsequent and/or simultaneous measurements of different quantum observables do not necessarily commute, making arbitrarily precise, joint measurements of those quantum variables impossible. For spin measurements, the Pauli-spin matrices are non-commuting, making arbitrarily precise measurement of a spin/qubit impossible; instead, only one component of the spin can be measured at the time, at the expense of loosing any information about the other components. This concept will be shown to be at the very basis of several quantum communication schemes (see Sect. 1.2), using non-orthogonal states (corresponding to non-commuting measurement operators) to encode quantum information.

1.1.2 No-Cloning Theorem

Besides the combination of coherent evolution and incoherent measurement, another crucial aspect of quantum bits can be derived from first quantum mechanical principles: the impossibility to copy an arbitrary quantum object. This is the basis of the famous *no-cloning* theorem of quantum mechanics [19], and is based on the linearity of quantum mechanics. The argument is as follows: suppose one has an arbitrary qubit, $|\Psi\rangle$, and an ancilla-qubit, $|\chi\rangle$ into which the state of $|\Psi\rangle$ needs to be copied (we assume, without loss of generality, that the initial state of the ancilla is always $|\downarrow\rangle$). Then, denoting the cloning operation as $C(|\Psi\rangle \otimes |\chi\rangle)$, we should have the following set of identities:

$$C(|\downarrow\rangle \otimes |\downarrow\rangle) = |\downarrow\rangle \otimes |\downarrow\rangle \quad (1.3)$$

$$C(|\uparrow\rangle \otimes |\downarrow\rangle) = |\uparrow\rangle \otimes |\uparrow\rangle \quad (1.4)$$

$$C([\alpha|\downarrow\rangle + \beta|\uparrow\rangle] \otimes |\downarrow\rangle) = \alpha C(|\downarrow\rangle \otimes |\downarrow\rangle) + \beta C(|\uparrow\rangle \otimes |\downarrow\rangle)$$
$$= \alpha |\downarrow\rangle \otimes |\downarrow\rangle + \beta |\uparrow\rangle \otimes |\uparrow\rangle \quad (1.5)$$
$$\neq [\alpha|\downarrow\rangle + \beta|\uparrow\rangle] \otimes [\alpha|\downarrow\rangle + \beta|\uparrow\rangle] \quad (1.6)$$

The discrepancy between Eq. 1.5, which is based on the linearity of quantum mechanical operations, and Eq. 1.6, which represents the target state of a true quantum mechanical copying device, is the proof by contradiction of the absence of such a cloning possibility. Obviously, this argument is only valid for arbitrary, unknown single quantum states: if the exact nature of the coherent superposition were known beforehand, or could be obtained through repeated measurement (e.g., if many copies of the unknown quantum state already exist), then operating the cloning device in an eigenbasis through coherent measurement basis rotation would still allow for copying of the quantum state.

1.1.3 Multiple Qubits: Non-classical Correlations

For multiple qubits, the joint Hilbert space contains states that cannot be written as the tensor-product of individual qubit states – in other words, where the state of one qubit is not independent from that of another. The characteristic correlations of such non-separable multi-qubit states are commonly referred to by the term *entanglement*, the English translation of the German word Verschränkung that was used by Erwin Schrödinger in the context of quantum mechanical correlations [20, 21].

The simplest entangled states involve two qubits; famous examples include the EPR-Bell states [22]. Let us consider one such state, the $|\Psi^-\rangle$-state, also known as the singlet state in quantum chemistry. With the axis conventions used in our Bloch-sphere description (Fig. 1.2), this state can be written as follows:

$$|\Psi^-\rangle = \frac{1}{\sqrt{2}}[|\uparrow\rangle_{1,z} \otimes |\downarrow\rangle_{2,z} - |\downarrow\rangle_{1,z} \otimes |\uparrow\rangle_{2,z}] \tag{1.7}$$

In Eq. 1.7, the subscripts refer to both the qubit (1,2) and the basis (here, z) used in the description. For a measurement in the z-basis, we immediately observe two things: on the one hand, the superposition of states reduces to a single state upon single-qubit measurement (either $|\uparrow\rangle_{1,z} \otimes |\downarrow\rangle_{2,z}$ or $|\downarrow\rangle_{1,z} \otimes |\uparrow\rangle_{2,z}$, each with 50 % probability); on the other hand, the resulting measured states show distinct (anti)correlations between the spins: regardless of whether the first spin is measured to be up or down, the other spin is always measured to be in the opposite state.

While such an anticorrelation could be observed classically in a statistical mixture of spins where only one of them can be in the up-(down-)state, the quantum mechanical correlations are much stronger than that. This can be observed by measuring the spins in another basis (we choose the x-basis, though the statement is valid for any other basis):

$$|\uparrow\rangle_z = \frac{1}{\sqrt{2}}[|\uparrow\rangle_x - |\downarrow\rangle_x] = \frac{1}{\sqrt{2}}[|\rightarrow\rangle_z - |\leftarrow\rangle_z] \tag{1.8}$$

$$|\downarrow\rangle_z = \frac{1}{\sqrt{2}}[|\uparrow\rangle_x + |\downarrow\rangle_x] = \frac{1}{\sqrt{2}}[|\rightarrow\rangle_z + |\leftarrow\rangle_z] \tag{1.9}$$

$$|\Psi^-\rangle = \frac{1}{\sqrt{8}}[(|\uparrow\rangle_{1,x} - |\downarrow\rangle_{1,x}) \otimes (|\uparrow\rangle_{2,x} + |\downarrow\rangle_{2,x}) - (|\uparrow\rangle_{1,x} + |\downarrow\rangle_{1,x}) \otimes (|\uparrow\rangle_{2,x} - |\downarrow\rangle_{2,x})]$$

$$= \frac{1}{\sqrt{2}}[|\uparrow\rangle_{1,x} \otimes |\downarrow\rangle_{2,x} - |\downarrow\rangle_{1,x} \otimes |\uparrow\rangle_{2,x}] \tag{1.10}$$

From Eq. 1.10, we immediately see that the spin-correlations exist in the x-basis as well, which cannot be explained by a classical or statistical mixture description.

Besides the aforementioned $|\Psi^-\rangle$-(singlet-)state, there are three other EPR-Bell states, the triplet-states:

$$|\Psi^+\rangle = \frac{1}{\sqrt{2}}[|\uparrow\rangle_{1,z} \otimes |\downarrow\rangle_{2,z} + |\downarrow\rangle_{1,z} \otimes |\uparrow\rangle_{2,z}] \quad (1.11)$$

$$|\Phi^+\rangle = \frac{1}{\sqrt{2}}[|\uparrow\rangle_{1,z} \otimes |\uparrow\rangle_{2,z} + |\downarrow\rangle_{1,z} \otimes |\downarrow\rangle_{2,z}] \quad (1.12)$$

$$|\Phi^-\rangle = \frac{1}{\sqrt{2}}[|\uparrow\rangle_{1,z} \otimes |\uparrow\rangle_{2,z} - |\downarrow\rangle_{1,z} \otimes |\downarrow\rangle_{2,z}] \quad (1.13)$$

These four EPR-Bell are maximally entangled 2-qubit states.[6] It can be shown [1, 22] that they form a complete and orthonormal basis for the Hilbert space of the 2-qubit states, and that therefore, by definition, any 2-qubit state can be decomposed into them. While a very powerful concept that shall be exploited further (see Sect. 1.1.3.1), the fragility of the EPR-Bell states to single-qubit measurement (collapse into pure, separable states) requires special care in dealing with them (see Sect. 1.1.3.2).

1.1.3.1 Entanglement as a Resource

The above description of entanglement did not require the qubits to reside in the same location – in fact, interesting applications arise for *remotely entangled* qubits, which are, among others, the subject of the famous Einstein-Podolsky-Rosen Gedankenexperiment [21] and many others that are at the very heart of the intersection of quantum (meta)physics.

From a heuristic and application-driven perspective, two particular consequences of remote entanglement shall turn out to be crucial for the realization of quantum repeaters: entanglement swapping, and entanglement distribution.

Entanglement swapping relies on the presence of several entangled pairs, and the possibility of realizing a full, joint Bell-state measurement on one of the qubits of each pair. Figure 1.3 illustrates the basic procedure. Suppose we start from 2, initially fully independent entangled pairs: $|\Psi^-_{1,2}\rangle$ and $|\Psi^-_{4,3}\rangle$. The singlet states are chosen for convenience, though similar results can be derived starting from any of the triplet states.

Insertion of the Bell-states of qubits 2 and 3 (which, as we recall, form a complete basis for the Hilbert-subspace of those two qubits), leads formally to the following result:

---

[6]We shall refrain, for now, from formal definitions and metrics of entanglement, and refer to Chap. 7 for more details. For now, it suffices to contrast the maximally entangled 2-qubit states with a state like, e.g. $0.1|\uparrow\rangle_{1,z} \otimes |\downarrow\rangle_{2,z} + 0.995|\downarrow\rangle_{1,z} \otimes |\uparrow\rangle_{2,z}$ – the latter, while entangled, is very close to the separable state $|\downarrow\rangle_{1,z} \otimes |\uparrow\rangle_{2,z}$. Handwaivingly, for now, we shall refer to the maximally entangled states as those 2-qubit states that are the furthest removed from any separable 2-qubit state.

1.1 On Quantum Bits, Their Measurement, and the Inability to Clone Them

Fig. 1.3 The canonical example of an entanglement-swapping scheme, starting from 2 initially unrelated singlet states. Bell refers to a full, joint Bell state measurement on the qubits 2 and 3

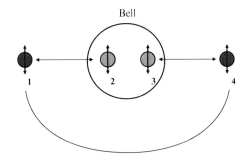

$$|\Psi\rangle = \left|\Psi^-_{1,2}\right\rangle \otimes \left|\Psi^-_{4,3}\right\rangle \tag{1.14}$$

$$= \frac{1}{2}[|\uparrow\rangle_{1,z} \otimes |\downarrow\rangle_{2,z} - |\downarrow\rangle_{1,z} \otimes |\uparrow\rangle_{2,z}] \otimes [|\uparrow\rangle_{4,z} \otimes |\downarrow\rangle_{3,z} - |\downarrow\rangle_{4,z} \otimes |\uparrow\rangle_{3,z}] \tag{1.15}$$

$$= \frac{1}{2}[\left|\Phi^+_{1,4}\right\rangle \otimes \left|\Phi^+_{2,3}\right\rangle + \left|\Psi^-_{1,4}\right\rangle \otimes \left|\Psi^-_{2,3}\right\rangle - \left|\Phi^-_{1,4}\right\rangle \otimes \left|\Phi^-_{2,3}\right\rangle - \left|\Psi^+_{1,4}\right\rangle \otimes \left|\Psi^+_{2,3}\right\rangle] \tag{1.16}$$

Equation 1.16, while formally a simple mathematical re-ordering and manipulation of Eq. 1.14, is very powerful: it clearly illustrates that, *if a joint Bell state measurement can be performed on qubits 2 and 3*, then the result of such a measurement is a new entangled state, now between qubits 1 and 4. Note that, again, we are not requiring qubits 1 and 4 to be in each other's vicinity, or even to directly interact with each other – rather, we require them both to be entangled to auxiliary qubits, 2 and 3, for which a joint Bell-state measurement should be feasible. Historically, such a scheme is referred to as entanglement swapping[7] [25].

A straightforward implementation, also suggested in Fig. 1.3, consists of two non-locally entangled pairs, with the spatial range of their entanglement respectively X and Y; entanglement swapping can then extend the range of the resulting entangled pair to the joint distance of X + Y – and so on, if multiple swapping schemes are nested together. This forms a basic ingredient of a quantum repeater [13, 14, 26]: a series of auxiliary entangled pairs, each stretching out over, say, 50 km, which are successively swapped in order to obtain truly long-range, possibly even intercontinental [27] entanglement.

Quantum teleportation is based on a very much similar approach. Here, as illustrated in Fig. 1.4, a single, unknown qubit $|\phi\rangle$ and an entangled pair $\left|\Psi^-_{2,3}\right\rangle$ are the basic ingredients. While the no-cloning theorem [19] prevents copying of

---

[7] In all of the present discussion, we assume to be working on pure entangled states, and to be able to perform perfect Bell measurements. In practice, imperfectly entangled states and/or statistical mixtures can be used in a distillation procedure known as *entanglement purification*: an initial set of partially entangled states can, under certain conditions, be transformed to a new, reduced set with improved entanglement properties. We refer to Refs. [23] and [24] for more details.

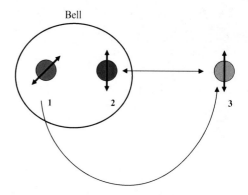

Fig. 1.4 The canonical example of an entanglement-based quantum-teleportation scheme, starting from an arbitrary, unknown qubit and an initially unrelated entangled pair. Bell refers to a full, joint Bell state measurement on the qubits 2 and 3

the arbitrary qubit state, it does not disallow a procedure where the qubit state is somehow absorbed, and re-created. Note that any such scheme where the state of the qubit were to be measured during the course of it, would automatically fail due to the collapse of the wavefunction upon measurement.

Again, insertion of the complete basis set of Bell states of qubits 2 and 3 leads, formally, to the following result:

$$|\chi\rangle_1 = \alpha|\uparrow\rangle_1 + \beta|\downarrow\rangle_1 \tag{1.17}$$

$$|\Psi\rangle = |\chi\rangle_1 \otimes |\Psi^-\rangle_{2,3}$$

$$= \frac{1}{2}[\alpha|\uparrow\rangle_1 + \beta|\downarrow\rangle_1] \otimes [|\uparrow\rangle_{2,z} \otimes |\downarrow\rangle_{3,z} - |\downarrow\rangle_{2,z} \otimes |\uparrow\rangle_{3,z}] \tag{1.18}$$

$$= |\Phi^+\rangle_{1,2} \otimes [\alpha|\downarrow\rangle_3 - \beta|\uparrow\rangle_3] + |\Phi^-\rangle_{1,2} \otimes [\alpha|\downarrow\rangle_3 + \beta|\uparrow\rangle_3]$$
$$- |\Psi^+\rangle_{1,2} \otimes [\alpha|\uparrow\rangle_3 - \beta|\downarrow\rangle_3] - |\Psi^-\rangle_{1,2} \otimes [\alpha|\uparrow\rangle_3 + \beta|\downarrow\rangle_3] \tag{1.19}$$

$$= |\Phi^+\rangle_{1,2} \otimes R_{\Phi^+}(|\chi\rangle_3) + |\Phi^-\rangle_{1,2} \otimes R_{\Phi^-}(|\chi\rangle_3)$$
$$- |\Psi^+\rangle_{1,2} \otimes R_{\Psi^+}(|\chi\rangle_3) - |\Psi^-\rangle_{1,2} \otimes R_{\Psi^-}(|\chi\rangle_3) \tag{1.20}$$

In Eq. 1.20, $R_{|\Psi^-\rangle}$ etc. refer to single qubit rotations on qubit 3, conditional on the joint Bell state measurement between qubits 1 and 2. In this form, the teleportation is obvious: depending on the Bell state measurement, a particular single-qubit operation needs to be applied to the third qubit, after which the third qubit becomes a perfect copy of the first one, *without ever being measured*. As 1 bit of classical information needs to be transferred from the location of qubits 1, 2 to 3, no superluminal communication is possible, and therefore no violation of special relativity occurs.

1.1.3.2 Bell State Measurements

As discussed above, both entanglement swapping and quantum teleportation depend critically on the presence of a joint Bell-state measurement between initially independent qubits. As the Bell states are explicitly superpositions of pairs of single-qubit states, any measurement scheme that would measure an individual qubit would therefore be incompatible with such a Bell state measurement.

A first, deterministic type of Bell-measurement involves the conditional interaction between two qubits in an entangling, 2-qubit gate. The canonical example is the CNOT-gate, where the spin of a target qubit is flipped, depending on the state of an ancilla-qubit [1]. Denoting the gate as CN, and using qubit 1 as the ancilla, 2 as the target, we obtain the following results:

$$CN(|\uparrow\rangle_1 \otimes |\uparrow\rangle_2) = |\uparrow\rangle_1 \otimes |\uparrow\rangle_2$$
$$CN(|\uparrow\rangle_1 \otimes |\downarrow\rangle_2) = |\uparrow\rangle_1 \otimes |\downarrow\rangle_2$$
$$CN(|\downarrow\rangle_1 \otimes |\uparrow\rangle_2) = |\downarrow\rangle_1 \otimes |\downarrow\rangle_2$$
$$CN(|\downarrow\rangle_1 \otimes |\downarrow\rangle_2) = |\downarrow\rangle_1 \otimes |\uparrow\rangle_2 \quad (1.21)$$

When combined with a single-qubit rotation on the ancilla qubit (e.g. a Hadamard gate H [1]), such a CNOT gate transforms an entangled, 2-qubit Bell-state into a separable 2-qubit state, after which measurements on the individual qubits can be performed:

$$CN(|\Psi^-\rangle) = CN(\frac{1}{\sqrt{2}}[|\uparrow\rangle_1 \otimes |\downarrow\rangle_2 - |\downarrow\rangle_1 \otimes |\uparrow\rangle_2])$$
$$= \frac{1}{\sqrt{2}}(|\uparrow\rangle_1 - |\downarrow\rangle_1) \otimes |\downarrow\rangle_2 \quad (1.22)$$
$$H_1(CN(|\Psi^-\rangle)) = |\downarrow\rangle_1 \otimes |\downarrow\rangle_2 \quad (1.23)$$

Similar results are valid for the other EPR-Bell states, each of which is mapped into a different, separable 2-qubit state after sequential application of the CNOT-gate and a 1-qubit Hadamard gate.

Physical implementations of CNOT gates have been realized in many systems, including but not limited to trapped ions, electron spin qubits, superconducting qubits, etc. [4]. For photons, however, the very weak mutual interaction limits the realization of such a scheme to materials with large Kerr non-linearities [28], and even there, the interaction strength is typically insufficient.

Another, probabilistic scheme exists however for Bell-state measurements, one that is based on beamsplitter-interference and therefore lends itself well for photonic implementations [29, 30]. Figure 1.5 illustrates the convention used in the description of the beamsplitter: ports a and b as inputs, and c and d as outputs.

Fig. 1.5 Schematic outline of a photon beamsplitter; a and b are the input ports, and c and d the outputs

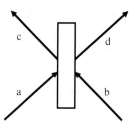

For single photon input states, say $|\Phi\rangle$ or $|\Psi\rangle$ at inputs a or b, the beamsplitter coherently mixes the inputs to yield the following results:

$$|\Phi\rangle_a \to \frac{1}{\sqrt{2}}[|\Phi\rangle_c + |\Phi\rangle_d] \tag{1.24}$$

$$|\Psi\rangle_b \to \frac{1}{\sqrt{2}}[|\Psi\rangle_c - |\Psi\rangle_d] \tag{1.25}$$

$$|\Phi\rangle_a \otimes |\Psi\rangle_b \to \frac{1}{2}[|\Phi\rangle_c + |\Phi\rangle_d] \otimes [|\Psi\rangle_c - |\Psi\rangle_d] \tag{1.26}$$

$$|\Phi\rangle_a \otimes |\Phi\rangle_b \to \frac{1}{\sqrt{2}}[|\Phi\rangle_c \otimes |\Phi\rangle_c - |\Phi\rangle_d |\Phi\rangle_d] \tag{1.27}$$

In Eq. 1.27, valid for *indistinguishable* single photons only, the destructive interference between the respective pathways from input a and b to the outputs results in bunching behavior for the photons at the output: both photons will emerge at the same output – the Hong-Ou-Mandel effect [31].

For two of the four EPR-Bell states ($|\Psi^{+,-}\rangle$), the HOM-effect for indistinguishable single photons gives rise to a unique output signature, that can be analyzed by means of single-photon detectors and other analyzers:

$$|\Psi^-\rangle_{a,b} \to \frac{1}{\sqrt{2}}[|\downarrow\rangle_c \otimes |\uparrow\rangle_d - |\uparrow\rangle_c \otimes |\downarrow\rangle_d] \tag{1.28}$$

$$|\Psi^+\rangle_{a,b} \to \frac{1}{\sqrt{2}}[|\uparrow\rangle_c \otimes |\downarrow\rangle_c - |\uparrow\rangle_d \otimes |\downarrow\rangle_d] \tag{1.29}$$

For single photons, the pseudo-spin can be encoded as a polarization state (H,V), the color of the photon, etc. For a $|\Psi^-\rangle_{a,b}$-state, both photons will emerge at different ports (see Eq. 1.28 – it can be shown that this is the only EPR-state for which this is the case), such that the detection of two photons at different outputs provides a unique signature for the presence of such a state. The $|\Psi^+\rangle_{a,b}$-state (Eq. 1.29) has both photons emerge at the same output, yet with opposite pseudospin, while the $|\Phi^{+,-}\rangle_{a,b}$-states can be shown to have the same pseudospin. Hence, a detection scheme based on polarizing beamsplitters (for polarization encoding) or dichroic mirrors (color encoding) and two detectors per output port will enable unambiguous detection of the $|\Psi^+\rangle_{a,b}$-state.

Given that this beamsplitter-based scheme is capable of detecting only half of the EPR-Bell states, it is in se probabilistic, with a 50 % probability of success, modulo photonic detection efficiencies. The scheme is particularly suited for the realization of the medium-range entanglement that is at the basis of entanglement swapping and quantum teleportation. Starting from two entangled pairs, e.g. entangled photon pairs from a parametric downconversion source [29] or two spin-photon entangled pairs (see Chap. 7), one can interfere one photon of each pair with one from the other pair, after first having it traverse a long distance in either free-space or low-loss optical fiber. Given the very weak intrinsic interaction strength of single photons, several tens to even hundreds of kilometers can be traversed this way, before probabilistic Bell-measurements project the other half of each pair into a joint, entangled state. Moreover, this projection is heralded: the observation of two 'clicks' of two particular beamsplitters reveals the realization of the projected entanglement, in principle with perfect accuracy.[8]

1.2 A Simple Quantum Communication Protocol

The (arguably) first practical application of QIP was in secure communication – quantum key distribution (QKD) for one-time-pad ciphers [12], as discussed before. The simplest possible QKD scheme was developed by Charles Bennett and Gilles Brassard in 1984 (BB84 [9]), and will be described at an elementary level below. It is essentially based on a combination of the no-cloning theorem for single qubits [19], as well as the Heisenberg uncertainty principle manifesting itself in collapse of the wavefunction along different, incompatible bases when using non-orthogonal states to encode information.

The essence of the scheme is indicated in Fig. 1.6. The counterparties, Alice and Bob, who want to share a secret key, each have access to a qubit measurement apparatus that they can use in different bases, corresponding to non-orthogonal eigenstates (say, $|\uparrow\rangle_z$, $|\downarrow\rangle_z$ or $|\uparrow\rangle_x$, $|\downarrow\rangle_x$). Alice now has a set of single qubits available, that she can measure *randomly* in either of these bases, and subsequently send to Bob – in practice, photons are used, and the original proposal mapped the qubit into photonic polarization states. Bob, in turn, measures the received qubits, again randomly choosing between bases. Up until this point, no information or key has been shared, and each party just has a table with measurement bases and results. Bob now announces his choice of bases to Alice, and vice versa: around 50 % of the

---

[8]The situation is slightly more complex when practical detectors are considered, which can have fake detection events known as dark counts. As long as the real detection events outnumber the dark counts significantly, reasonably high fidelity entanglement can be realized, and potentially further purified [24]. However, once the dark counts start swamping the real detection events, this scheme fails completely. We will discuss this in more detail in the next section.

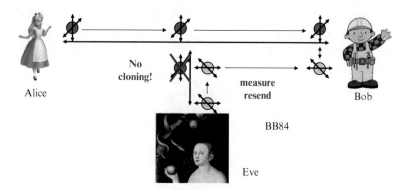

Fig. 1.6 Schematic of the BB84 QKD protocol. Alice and Bob attempt to share a secret key, which Eve tries to intercept. As she cannot clone the qubits used, she attempts to measure and resend them, which leads to detectable errors upon comparing Alice and Bob's code tables

cases should result in a correspondence between the chosen measurement bases, and can now be used: a *sifted key* has been established. Only Alice and Bob know which of the two possible measurement results appeared for the events with corresponding bases, and have thus established a secret key.

In the presence of an eavesdropper, the no-cloning theorem precludes simple copying of the qubits and then overhearing which qubits to look at. Instead, Eve has to resort to measuring the qubits, and then resending them. However, as she had no a priori information about which basis Alice or Bob are using (the assumption of perfectly random choices is critical here), she can only guess the measurement basis correctly with 50 % probability. In the other 50 %, she chooses the wrong basis, and collapses the qubit into a non-orthogonal eigenstate compared to the originally transmitted one. For the pseudo-spin case we described above (or a polarization mapping thereof), this now resent qubit is measured by Bob. In the comparison stage of the algorithm, when Alice and Bob establish corresponding measurement bases, he has again a 50 % probability of obtaining the same result as Alice: a qubit in, say, the x-basis, $|\uparrow\rangle_x$, when measured in the z-basis, has 50 % probability of ending up in either the $|\uparrow\rangle_z$- or $|\downarrow\rangle_z$-state. By comparing their results for a random subset of the corresponding measurements, Alice and Bob can establish an error rate. If Eve overhears every single qubit and resends it, she should project a 25 % error rate onto the key which Alice and Bob intend to share, the signature of which will instruct them to abandon their mission. For lower-rate, randomly chosen intercepts, the amount of information Eve can obtain about the key decreases, and Alice and Bob will notice a lower error rate – after which they can resort to classical cryptographic tricks such as privacy amplification to increase the security of their shared key [7, 8].

1.2.1 Practical Issues: Losses, Detectors and Such

It would go beyond the scope of this work to describe all the possible weaknesses in practical BB84 implementations; we shall instead focus on a few major ones, that both limit its possible use and have led to different, more elaborate schemes to become more prevalent for QKD.

As the BB84 scheme relies on the no-cloning theorem, its security depends on the quality of the single-photon sources used. While highly attenuated lasers can mimic some aspects of true single-photon sources, the small yet finite probability of having more than one photon per experimental event (intended bit sent) might allow Eve to circumvent the no-cloning theorem and obtain some information about the keys that remains undetected. In particular, if she were able to perform a non-destructive measurement of the number of photons per pulse, she could decide to only focus on those events with multiple photons, measure one of these, and Bob would be unable to detect any errors due to this attack strategy. There has been considerable success in realizing true, pulsed, single-photon sources [32, 33], though even those have non-zero probabilities of emitting more than one photon per pulse. In practice, more elaborate schemes can be used to deal with the multiple-photon emission, e.g. by relying on photon-number resolving detectors – at the expense of over-all performance [34].

A more serious issue is caused by the combination of qubit-loss and imperfect detectors. While, in the above scheme, Alice and Bob effectively post-select for cases where both of them observe a detector click, the rate of such coincidences obviously decreases with increased system losses. For photons, the losses are either due to absorption in optical fibers (some 0.2 dB/km for state-of-the-art telecom fiber), or by diffraction and simple lateral spreading out of the single photons when used, unguided, in free space. On top of that, the detectors used will sporadically detect spurious events (dark counts), unrelated to any real detected photon. Such dark counts will therefore result in fake signals and errors in the sifted key. Once the real count rate, reduced due to absorption/losses, falls below a certain threshold, the dark count signal will become large enough that no privacy amplification can be of any help: the scheme fails. For practical detectors, with Hz-level dark count rates, and fiber-optic communication, this will practically limit the BB84-scheme and its derivatives to distances of several 100 km [7, 8, 35].

Finally, a polarization-encoding of the qubits is undesirable due to inherent and constantly fluctuating birefringence in optical fibers due to small amounts of strain or differential thermal expansion. It is in principle possible to compensate these by applying test signals, though those would compete with real signals for bandwidth and therefore reduce the possible key generation rate. The use of polarization-maintaining fiber on the other hand is incompatible with the use of non-orthogonal basis-states. The polarization-encoded BB84 scheme is therefore limited to free-space applications, while fiber-based systems rely on BB84 variants such as differential-phase-shift (DPS) QKD [36, 37].

1.3 An Entanglement-Based Quantum Communication Protocol

A different QKD scheme, based on remote entanglement, was proposed by Arthur Ekert in 1991 [10] and slightly modified by Bennett, Brassard and Mermin the following year [11]. The latter authors also pointed to the obvious similarities with the existing BB84 protocol. In their approach, Alice and Bob each share one qubit of an entangled (singlet) EPR-Bell pair. They then, as in BB84, each randomly choose their measurement basis (x,z as in our convention before), and compare post-factum the choice of basis states. For the same choice of measurement basis, the perfect anticorrelation present in the singlet states guarantees that they both have exactly opposite results, which with a trivial inversion leads to an identical key. In contrast to the BB84 scheme, Alice does now not anymore 'choose' her qubits: both she and Bob accept the random measurement results that follow from collapse of the entangled EPR-Bell state, and obtain a sifted key from this process (Fig. 1.7).

Similar to the case of BB84, Eve cannot perform copies on the individual qubits, nor can she perform a single-qubit intercept-and-resend attack, as that will again result in errors upon comparison of parts of the sifted key by Alice and Bob. Moreover, Bennett and co-workers also demonstrated that Eve would not be able to fool Alice and Bob either by actively providing them with an entangled pair over which she would have any sort of control – where she would somehow be able to infer information about the key without being noticed. The argument goes as follows: for Eve to obtain information over the joint spin-state that Alice and Bob receive, she would need to create an (at least) 3-particle entangled state:

$$|\Phi\rangle = |\uparrow\uparrow\rangle_{A,B} \otimes |A\rangle_E + |\downarrow\downarrow\rangle_{A,B} \otimes |B\rangle_E + |\uparrow\downarrow\rangle_{A,B} \otimes |C\rangle_E + |\downarrow\uparrow\rangle_{A,B} \otimes |D\rangle_E \quad (1.30)$$

However, as Alice and Bob must not notice her tampering with the singlet EPR-Bell pair, this can only be achieved in the following way, where indeed perfect anticorrelations are obtained whenever Alice and Bob perform and compare their spin measurements in the same basis:

$$|\Phi\rangle = (|\uparrow\downarrow\rangle_{A,B} - |\downarrow\uparrow\rangle_{A,B}) \otimes |C\rangle_E \quad (1.31)$$

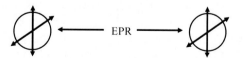

Fig. 1.7 Schematic of the BBM92 QKD protocol. Alice and Bob each receive one qubit of an entangled EPR-Bell pair, and perform random measurements in non-orthogonal bases. After basis-comparison, the sifted key is checked for errors, which would indicated the presence of an eavesdropper. The circles refer to the equator of the Poincaré-sphere that can be used for a polarization-implementation. For a detailed review, see Ref. [8]

Yet, in Eq. 1.31, Eve's quantum state is totally unentangled with the EPR-Bell pair that she provides to Alice and Bob, and will therefore not help her obtain any information about the whereabouts of Alice and Bob's qubits.

Provided that entangled EPR-Bell pairs can be obtained and used as a resource, the BBM92 scheme does not suffer from the same loss- and dark count limitations as the BB84 scheme; in some sense, it post-selects for those events where successful entanglement generation can be used for BB84-like information and key-sharing. Obviously, the establishment of such entangled pairs is the key assumption here: for long-distance entanglement, this will most likely rely on a photonic scheme, where qubit-light entanglement is used in a probabilistic, HOM-like interference scheme in order to swap the entanglement to the target qubits (see Sects. 1.1.3.1 and 1.1.3.2 for more details). Crucially, this procedure is heralded: a double-click event on the detectors indicates the successful realization of remote entanglement; its probability of success, however, scales badly with distance due to the combined effects of linear loss and dark counts. While photonic losses and detector dark counts do therefore limit the range of any single link that can be established this way, neighboring links can be used in further entanglement-swapping schemes to extend this range. If combined with long-lived qubits as quantum memories [26], successfully realized entangled pairs can be stored until all the other links have been obtained. Using this effective parallelism, the photonic loss and dark count restrictions on the length of single link can be circumvented: this is the very essence of a quantum repeater [13, 14, 26, 38].

1.3.1 Entanglement: Quantum One-Time Pad?

The potential use of entangled singlet EPR-Bell pairs in a BBM92 scheme suggests another, potentially more straightforward approach to quantum communication: instead of sacrificing the entangled pairs for the generation of quantum mechanically secure keys, it might be possible to use quantum mechanics to transport the message as a whole, in a quantum version of the one-time pad [8]. Provided that faithful singlet pairs are available (which could be tested through the same comparison scheme as used in BBM92 and described above, revealing the presence of eavesdroppers through excessive error rates), the message could be quantum teleported [39] as a whole. As the eavesdropper does not have access to the other half of the entangled pair, he or she cannot reproduce the original message after overhearing the classical bits indicating the quantum operations to be performed by the receiver. Only the faithful recipient of the original singlet EPR-Bell pairs, who does have access to the 'other half' will have be able to use this information.

In general, whether a quantum teleportation or a BBM92 scheme is used is a matter of taste and application dependent. Regardless, the presence of entangled EPR-Bell states allows for long-distance quantum communication, and its range is, unlike in the single-photon BB84-related schemes, extendable through repeated entanglement swapping.

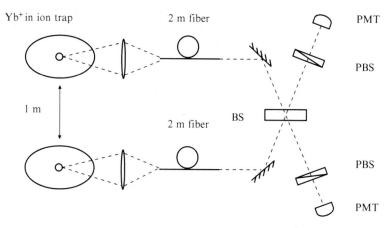

Fig. 1.8 Schematic of the first ion quantum teleportation experiment. The quantum state of one Yb^+ ion is mapped into a remote one, by means of ion-photon entanglement and a probabilistic Bell state measurement on the ion-entangled photon. *(P)BS* (polarizing) beamsplitter, *PMT* photo-multiplying tube (a type of single-photon detector). For more information, see Ref. [43]

1.3.2 Practical Implementation: Ion Traps

While the first entanglement swapping and quantum teleportation experiments involved photonic qubits and parametrically downconverted photonic EPR-Bell pairs [29, 40], the combination with long-lived, memory qubits requires different technologies. Trapped ions, arguably the most advanced of the matter qubits [4], and certainly among the longest-living ones, were used to demonstrate, successively, spin-photon entanglement [41], remote qubit entanglement through effective entanglement swapping after probabilistic, HOM-interference-based photonic Bell-state measurements [42] and quantum teleportation between remote ionic (Yb^+) qubits [43].

The latter scheme is illustrated in Fig. 1.8, and is a slight simplification of the canonical schemes described before (establishing remote qubit entanglement; Bell state measurement between the unknown qubit and one half of the EPR-Bell state), as it involves only two Yb^+-qubits, not three. After appropriate initialization of those qubits into respectively the arbitrary, to-be-teleported qubit state A ($\alpha |\downarrow\rangle_A + \beta |\uparrow\rangle_A$) and a coherent superposition for qubit B, ion-photon entanglement is created (see Chap. 7). With the particular initializations used, the effective, probabilistic two-photon Bell state measurement projects the two ionic qubits into a particular entangled state: $\alpha(|\downarrow\rangle_A + |\uparrow\rangle_A) \otimes |\uparrow\rangle_B - \beta(-|\downarrow\rangle_A + |\uparrow\rangle_A) \otimes |\downarrow\rangle_B$. Measurement of qubit A then projects qubit B into a measurement-outcome dependent superposition, that can be coherently rotated into the desired state.

1.4 Solid-State Based Quantum Repeaters

The realization that truly long-distance entanglement could be used for quantum communication in either BBM92-based schemes or through quantum teleportation, led to various proposals for generating such long-distance entanglement [13, 14, 26, 27]. Invariably, they are based on a form of nested entanglement purification and swapping, in the presence of long-lived quantum memories: the quantum repeater.

A basic outline of a quantum repeater is shown in Fig. 1.9. In a first step (a), a massively parallel series of entangled qubit pairs is realized. Probabilistic, heralded schemes based on HOM-like interference of qubit-entangled photons could be used for this. However, their probability of success is limited and critically dependent on the distance between the pairs due to exponential increase in the linear losses with growing distance. An optimal strategy therefore seems to consist of spacing the nodes several 10 km apart, and to apply a repeat-until-success strategy: once a pair is established, its state is kept, the qubits are untouched, and the EPR-Bell state is assumed to be maintained by virtue of the quantum memory inherent in the qubits.[9] This way, all links can be entangled in parallel, with the total required time approximately equal to the average time needed for a single link to be established.

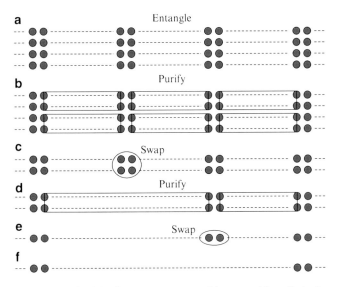

Fig. 1.9 Basic operation principle of a quantum repeater. Memory qubits at limited-range intervals are entangled, after which a series of nested entanglement-purification [23, 24] and entanglement-swapping [25] procedures are used to implement higher fidelity, longer-distance entanglement

---

[9]In practice, single-qubit memory times or coherences may be limited, requiring repeated 'refreshing' or correction of the memory: this can be realized by quantum error correction, which is an entire subject in itself, and one that has been extensively studied in the context of quantum computing – we refer to Refs. [1] and [4] for more details.

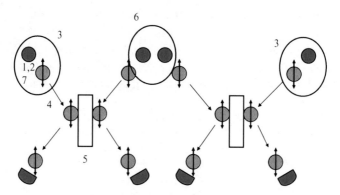

Fig. 1.10 Basic ingredients for a quantum repeater, as discussed in this work. *Green circles*: memory (spin) qubits; *orange circles*: single photonic qubits; *black-and-white rectangles*: beam-splitters for HOM-measurement; *green boxes*: single-photon detectors; *black-and-white circles*: entanglement operations

In a second step (b), the entangled pairs obtained in (a) are combined into a new, reduced set of entangled pairs in a process known as entanglement purification or distillation [23, 24]. As all quantum operations are imperfect, none of the pre-established entangled pairs would have perfect fidelity. Entanglement distillation starts from these imperfectly entangled pairs, and combines 2-qubit, entangling gates with single-qubit operations to obtain higher-fidelity pairs. This process can, again, be performed in parallel, on entire series of entangled pairs at once. The number of pairs required depends critically on the initial degree of entanglement, which in itself depends on the distance between the pairs, the fidelity of the quantum operations, the losses in the channels, the detector dark counts, etc.

When sufficiently entangled, distilled pairs are obtained, entanglement swapping can be used in a nested procedure (steps (c) and (e)), possibly combined with other, intermediate purification steps (d) that compensate for errors in previous swapping steps. The result is a longer-distance, high-fidelity entangled EPR-Bell pair, that could be used for secure quantum communication.

It is the potential for massive parallelism and high operation speed (see Chaps. 2 and 3 for more details) that make solid-state quantum repeaters an important yet challenging goal within the QIP-community at large. The work presented in this thesis falls within this framework.

1.4.1 Solid-State Quantum Repeaters: A Checklist

Figure 1.10 illustrates the basic quantum repeater ingredients as outlined before. They can be summarized as follows:

1. High-fidelity, coherent single qubit control: progress towards this goal will be presented in Chaps. 2–6;

2. Long-lived quantum bits (coherences), to be used as quantum memories: experimental studies and improvements are reported in Chaps. 3 and 6;
3. Spin-photon entanglement/interface: a coherent mapping of spin coherence to photonic coherence, in order to entangle stationary qubits with flying qubits. An experimental realization hereof is reported in Chap. 7;
4. Low-loss propagation of the photonic qubit, in order to transfer the quantum information over long distances. For fiber-based schemes, this requires photonic qubits at low-loss wavelengths (1,550 nm). An experimental implementation is reported in Chap. 7;
5. High-visibility photonic quantum interference at low-loss wavelengths: an HOM-based, effective Bell state measurement (probabilistic) to transfer spin-photon entanglement into spin-spin entanglement; ongoing work;
6. High-fidelity, fast, entangling 2-qubit gate for the stationary qubits: necessary for quantum teleportation and entanglement swapping; ongoing work;
7. High-fidelity, efficient quantum memory readout: ideally single-shot, and non-demolition; ongoing work

References

1. M. A. Nielsen and I. L. Chuang. *Quantum Computation and Quantum Information*. Cambridge University Press, 2000.
2. R. P. Feynman. Simulating physics with computers. *International Journal of Theoretical Physics*, 21:467, 1982.
3. S. Lloyd. Universal quantum simulators. *Science*, 273:1073, 1996.
4. T. D. Ladd, F. Jelezko, R. Laflamme, Y. Nakamura, C. Monroe, and J. L. O'Brien. Quantum computers. *Nature*, 464:45, 2010.
5. P. W. Shor. Algorithms for quantum computation: Discrete logarithms and factoring. *Proc. 35nd Annual Symposium on Foundations of Computer Science (Shafi Goldwasser, ed.)*, page 124, 1994.
6. L. K. Grover. From Schrödinger's equation to quantum search algorithm. *American Journal of Physics*, 69:769, 2001.
7. N. Gisin and R. Thew. Quantum communication. *Nat. Photonics*, 1:165, 2007.
8. N. Gisin, G. Ribordy, W. Tittel, and H. Zbinden. Quantum cryptography. *Rev. Mod. Phys.*, 74:145, 2002.
9. C. H. Bennett and G. Brassard. Quantum cryptography: Public key distribution and coin tossing. *Proceedings of the IEEE International Conference on Computers, Systems, and Signal Processing*, Bangalore:175, 1984.
10. A. K. Ekert. Quantum Cryptography Based on Bell's Theorem. *Phys. Rev. Lett.*, 67:661, 1991.
11. C. H. Bennett, G. Brassard, and N. D. Mermin. Quantum Cryptography without Bell's Theorem. *Phys. Rev. Lett.*, 68:557, 1992.
12. C. Shannon. Communication theory of secrecy systems. *Bell System Technical Journal*, 28:656, 1949.
13. H.-J. Briegel, W. Dür, J. I. Cirac, and P. Zoller. Quantum repeaters: The role of imperfect local operations in quantum communication. *Phys. Rev. Lett.*, 81:5932, 1998.
14. W. Dür, H.-J. Briegel, J. I. Cirac, and P. Zoller. Quantum repeaters based on entanglement purifcation. *Phys. Rev. A*, 59:170, 1999.

15. R. P. Feynman, F. L. Vernon, and R. W. Hellwarth. Geometrical representation of the schrodinger equation for solving maser problems. *J. Appl. Phys.*, 28:49, 1957.
16. L. Allen and J. H. Eberly. *Optical Resonance and Two-level Atoms*. Dover books on Physics, 1987.
17. W. Heisenberg. Über den anschaulichen Inhalt der quantentheoretischen Kinematik und Mechanik. *Zeitschrift für Physik*, 43:172, 1927.
18. J. von Neumann. *Mathematical Foundations of Quantum Mechanics*. Princeton University Press, 1955.
19. W. K. Wootters and W. H. Zurek. A single quantum cannot be cloned. *Nature*, 299:802, 1982.
20. E. Schrödinger. Die gegenwärtige Situation in der Quantenmechanik. *die Naturwissenschaften*, 23:807, 1935.
21. A. Einstein, B. Podolsky, and N. Rosen. Can Quantum-Mechanical Description of Physical Reality be Considered Complete? *Physical Review*, 47:777, 1935.
22. J. S. Bell. *Speakable and Unspeakable in Quantum Mechanics*. Cambridge University Press, 1987.
23. C. H. Bennett, H. J. Bernstein, S. Popescu, and B. Schumacher. Concentrating Partial Entanglement by Local Operations. *Phys. Rev. A*, 53:2046, 1996.
24. C. H. Bennett, G. Brassard, B. Popescu, S.and Schumacher, J. A. Smolin, and W. K. Wooters. Purification of Noisy Entanglement and Faithful Teleportation via Noisy Channels. *Phys. Rev. Lett.*, 76:722, 1996.
25. M. Zukowski, A. Zeilinger, M. A. Horne, and A. K. Ekert. Event-Ready-Detectors Bell Experiment via Entanglement Swapping. *Phys. Rev. Lett.*, 71:4287, 1993.
26. N. Sangouard, C. Simon, H. de Riedmatten, and N. Gisin. Quantum repeaters based on atomic ensembles and linear optics. *Rev. Mod. Phys.*, 83:33, 2011.
27. H. J. Kimble. The quantum internet. *Nature*, 453:1023, 2008.
28. N. Imoto, H. A. Haus, and Y. Yamamoto. Quantum nondemolition measurement of the photon number via the optical Kerr effect. *Phys. Rev. A*, 32:2287, 1985.
29. J.-W. Pan, D. Bouwmeester, H. Weinfurter, and A. Zeilinger. Experimental Entanglement Swapping: Entangling Photons That Never Interacted. *Phys. Rev. Lett.*, 80:3891, 1998.
30. E. Knill, R. Laflamme, and G. J. Milburn. A scheme for efficient quantum computation with linear optics. *Nature*, 409:46, 2001.
31. C. K. Hong, Z. Y. Ou, and Mandel L. Measurement of Subpicosecond Time Intervals between Two Photons by Interference. *Phys. Rev. Lett.*, 59:2044, 1987.
32. M. Pelton, C. Santori, J. Vuckovic, B. Zhang, G. S. Solomon, J. Plant, and Y. Yamamoto. Efficient source of single photons: A single quantum dot in a micropost microcavity. *Phys. Rev. Lett.*, 89:233602, 2002.
33. C. Santori, M. Pelton, G. Solomon, Y. Dale, and Y. Yamamoto. Triggered single photons from a quantum dot. *Phys. Rev. Lett.*, 86:1502, 2001.
34. E. Waks, K. Inoue, S. Santori, D. Fattal, J. Vuckovic, G. S. Solomon, and Y. Yamamoto. Quantum cryptography with a photon turnstile. *Nature*, 420:762, 2002.
35. H. Takesue, S. W. Nam, Q. Zhang, R. H. Hadfield, T. Honjo, K. Tamaki, and Y. Yamamoto. Quantum key distribution over a 40-dB channel loss using superconducting single-photon detectors. *Nat. Photonics*, 1:343, 2007.
36. K. Inoue, E. Waks, and Y. Yamamoto. Differential phase shift quantum key distribution. *Phys. Rev. Lett.*, 89:037902, 2002.
37. K. Inoue, E. Waks, and Y. Yamamoto. Differential-phase-shift quantum key distribution using coherent light. *Phys. Rev. A*, 68:022317, 2003.
38. Z.-S. Yuan, Y.-A. Chen, B. Zhao, S. Chen, J. Schmiedmayer, and J.-W. Pan. Experimental demonstration of a BDCZ quantum repeater node. *Nature*, 454:1098, 2008.
39. C. H. Bennett, G. Brassard, C. Crépeau, R. Jozsa, A. Peres, and W. K. Wootters. Teleporting an unknown quantum state via dual classical and einstein-podolsky-rosen channels. *Phys. Rev. Lett.*, 70:1895, 1993.
40. D. Bouwmeester, J.-W. Pan, K. Mattle, M. Eibl, H. Weinfurter, and A. Zeilinger. Experimental quantum teleportation. *Nature*, 390:575, 1997.

References

41. B. B. Blinov, D. L. Moehring, L.-M. Duan, and C. Monroe. Observation of entanglement between a single trapped atom and a single photon. *Nature*, 428:153, 2004.
42. D. L. Moehring *et al*. Entanglement of single-atom quantum bits at a distance. *Nature*, 449:68, 2007.
43. S. Olmschenk, D. N. Matsukevich, P. Maunz, D. Hayes, L.-M. Duan, and C. Monroe. Quantum Teleportation Between Distant Matter Qubits. *Science*, 323:486, 2009.

Chapter 2
Quantum Memories: Quantum Dot Spin Qubits

The quantum bits used in the remainder of this work, are individual electron (Chaps. 3–5 and 7) or hole spins (Chap. 6) in self-assembled quantum dots [1, 2]. Spins, either as direct spin-1/2 particles or as pseudospin-submanifolds of larger systems, are generally considered as good candidate-qubits due to their relatively limited interaction with the environment [3]. In addition, and as we shall show below, the confinement of the spin to quantum dots provides an additional protection of the spin degree of freedom.

'Quantum dot' is a generic name for a plethora of different, effectively 0-dimensional structures, where quantum effects such as quantum confinement actively affect the behavior of said structures. Over the last decade, so-called electrically defined quantum dots [2, 4] have gained a lot of attention: these are structures, consisting of a two-dimensional electron gas (2-DEG), where electrical contacts define small islands withing the 2-DEG, to the extent that quantum confinement and mutual interactions significantly restrict the number of allowed states within the quantum dot (few- to single-electron regime).

The quantum dots in the present work are of a different nature: these are self-assembled semiconductor quantum dots, MBE-grown through the Stranski-Krastanov method [1, 5]. A schematic outline is shown in Fig. 2.1. Due to the build-up of excessive strain during the MBE layer growth, small, nm-scale three-dimensional islands of semiconductor material coalesce, with a different composition and (often) lattice constant than the host matrix, thereby relieving part of the strain. The difference in composition, and the respective band-line-up in the conduction and valence band, can create a structure where electrons and/or holes are effectively bound to the quantum dot due to the band-discontinuity with the host material.

Due to the nm-size scale of the quantum dots (see Fig. 2.1a for an SEM micrograph of uncapped quantum dots), quantization and inter-particle interactions reduce the number of available states inside the quantum dots, which are typically in the few-electron (hole) regime, similar to the gate-defined quantum dots. However, unlike the gate-defined quantum dots, the three-dimensional, band-discontinuity

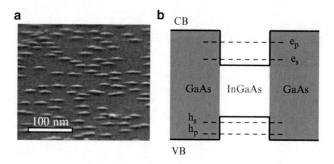

Fig. 2.1 Self-assembled quantum dots: SEM-micrograph of uncapped dots (**a**) and schematic outline of the band- and level structure (**b**). *CB* conduction band, *VB* valence band, $e_{s,p}$ s- and p-shell electron states, $h_{s,p}$ s- and p-shell hole states

induced confinement can give rise to true 3D confinement for both electrons and holes [1, 5]. When composed of optically active, direct bandgap semiconductors, such self-assembled quantum dots can therefore bind complexes of electrons and holes: excitons, which can have large optical dipole moments and result in fast optical recombination. For the remainder of this work, we shall focus on one particular material system: the InAs-GaAs system, where the optically active quantum dots consist of In(Ga)As islands within a GaAs host matrix. As we can see in Fig. 2.1a, where the top GaAs layer was removed, such quantum dots have a pancake-like shape, with some 2–4 nm height and between 20 and 50 nm in diameter.

The large confinement energy and reduced density of states provides an excellent protection for the spin states of individual charge carriers inside the quantum dot [5]. While the magnetic dipole moment of a single electron spin is very small [5], relativistic effects give rise to spin-orbit interactions that couple the spin to the particle momentum. In bulk, such coupling can easily give rise to spin flips due to different types of scattering events. The reduced density of states in quantum dots makes such spin flip processes much less effective, thereby providing additional protection for the spin qubits [6–8].

2.1 Quantum Dot Level Structure

Using effective-mass theory [5], the effect of the lattice on a single charged particle confined within the quantum dot can be decoupled from the effects of the confinement due to band-edge discontinuities. For a single quantum dot electron or hole, this allows for a separation of variables approach to the solution of the Schrödinger equation:

$$H_{total}|\Psi\rangle_{total} = E_{total}|\Psi\rangle_{total} \tag{2.1}$$

2.1 Quantum Dot Level Structure

$$|\Psi\rangle_{total} = |\Phi\rangle_{QD}|\chi\rangle_{band} \tag{2.2}$$

$$(H_{band} + H_{QD})|\Phi\rangle_{QD}|\chi\rangle_{band} = (E_{band} + E_{QD})|\Phi\rangle_{QD}|\chi\rangle_{band} \tag{2.3}$$

$$H_{QD}|\Phi\rangle_{QD} = E_{QD}|\Phi\rangle_{QD}$$

$$H_{QD} = \frac{-\hbar^2 \nabla^2}{2m_{\text{eff}}} + U_{QD} \tag{2.4}$$

Equation 2.4 for the envelope wavefunction $|\Phi\rangle_{QD}$ that stretches out over the entire quantum dot is formally equivalent to that of a particle in a three-dimensional box with U_{QD} representing the effect of the band discontinuity and any other extrinsic potentials. It describes the effects of the quantum confinement within a band inside the quantum dot, characterized by its effective mass m_{eff}.[1] In view of the different size scales of the self-assembled dots (vertical confinement: 2–4 nm, horizontal: 20–50 nm), the dominant quantization axis is along the growth direction. In analogy with the simple, one-dimensional particle-in-a-box model, an energy level structure develops, with s- and p-type envelope functions alternating. Figure 2.1 indicates the s- and p-shell energy levels for the electron and lowest-lying hole band respectively.

In Eq. 2.2, $|\chi\rangle_{band}$ reflects the symmetry of the underlying lattice. Using $\vec{k} \cdot \vec{p}$ perturbation theory for the zinc-blende semiconductors [9] (such as GaAs and InAs), this band-related wavefunction can be shown to have s-like symmetry for electrons, and p-like symmetry for holes [5]. Moreover, and again due to the symmetry of the underlying lattice, the electron wavefunction is singly degenerate (doubly when accounting for spin), and the hole wavefunction triply degenerate (six-fold when accounting for spin). Due to the coupling of the spin and the orbital angular momentum, the spin and orbit-part of the wavefunction hybridize. In combination with the confinement-induced splitting (the effective mass of each of the three bands is different, leading to different quantization energies along the growth direction), we obtain three spin-degenerate subbands: he heavy hole (HH), light hole (LH) and split-off (SO) bands. Using the convention used in Fig. 2.4b, with the growth and dominant confinement axis along the z-direction, and a potential magnetic field along x, we can describe the resulting wavefunctions in a hydrogen-atom-like fashion [5]:

$$|\chi\rangle_{e,\uparrow} = i|S\rangle \otimes |\uparrow\rangle_z \tag{2.5}$$

$$|\chi\rangle_{e,\downarrow} = i|S\rangle \otimes |\downarrow\rangle_z \tag{2.6}$$

---

[1] Strictly speaking, this is not exactly correct: in view of the band discontinuity, this Hamiltonian is not Hermitian. Several approaches have been developed to cope with this problem, among them the BenDaniel-Duke solution for one-dimensional discontinuities: $H_{BDD} = \frac{-\hbar^2}{2}\frac{\partial}{\partial z}\left(\frac{1}{m_{eff}(z)}\frac{\partial}{\partial z}\right) + U(z)$ [5].

$$|\chi\rangle_{HH,\uparrow} = |\Uparrow\rangle_z = -i\frac{1}{\sqrt{2}}(|X\rangle + i|Y\rangle) \otimes |\uparrow\rangle_z \qquad (2.7)$$

$$|\chi\rangle_{HH,\downarrow} = |\Downarrow\rangle_z = i\frac{1}{\sqrt{2}}(|X\rangle - i|Y\rangle) \otimes |\downarrow\rangle_z \qquad (2.8)$$

$$|\chi\rangle_{LH,\uparrow} = -i\frac{1}{\sqrt{6}}(|X\rangle + i|Y\rangle) \otimes |\downarrow\rangle_z + i\sqrt{\frac{2}{3}}|Z\rangle \otimes |\uparrow\rangle_z \qquad (2.9)$$

$$|\chi\rangle_{LH,\downarrow} = +i\frac{1}{\sqrt{6}}(|X\rangle - i|Y\rangle) \otimes |\uparrow\rangle_z - i\sqrt{\frac{2}{3}}|Z\rangle \otimes |\downarrow\rangle_z \qquad (2.10)$$

$$|\chi\rangle_{SO,\uparrow} = -i\sqrt{\frac{1}{3}}(|X\rangle + i|Y\rangle) \otimes |\downarrow\rangle_z - i\sqrt{\frac{1}{3}}|Z\rangle \otimes |\uparrow\rangle_z \qquad (2.11)$$

$$|\chi\rangle_{SO,\downarrow} = i\sqrt{\frac{1}{3}}(|X\rangle - i|Y\rangle) \otimes |\uparrow\rangle_z - i\sqrt{\frac{1}{3}}|Z\rangle \otimes |\downarrow\rangle_z \qquad (2.12)$$

As the effective mass of the different hole sub-bands is indeed different, the strong z-confinement will split the heavy holes from the light holes and split-off band by several 10 meV for typical InGaAs quantum dots. For most low-temperature experiments (few K to mK, He^4 or He^3-He^4 (dilution) refrigeration temperatures), this allows us to ignore any but the heavy holes. Hence, in the remainder of this work, we shall focus on four particular states: the s-shell electron states (with spin-up or spin-down), and the s-shell heavy holes. The latter are technically a submanifold of the $J = 3/2$ manifold, but the splitting off of the light holes allows a description in terms of an effective two-level system and therefore pseudo-spin-1/2 – we refer to Chap. 6 for an explicit demonstration hereof.

While we will focus on the lowest-lying electron and hole spin states for use as qubits, the excited states can be used for optically addressing the quantum dots. A simple schematic is indicated in Fig. 2.2: while the lowest-lying electron and hole

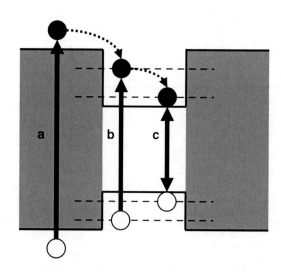

Fig. 2.2 Schematic overview of different excitation and recombination processes in self-assembled quantum dots. *Solid arrows*: optical processes; *dashed arrows*: non-radiative processes. (**a**) Above-band excitation; (**b**) quasi-resonant excitation; (**c**) resonant excitation

states in the quantum dot can only relax radiatively at low temperatures, with some 1 ns interband radiative recombination times [10–14], the excited states can decay within their respective bands non-radiatively due to e.g. phonon-emission, on much faster timescales. Besides the resonant excitation of electron-hole pairs (excitons, process (c) in Fig. 2.2), these excited states mediate both above-band excitation (process (a)) with subsequent non-radiative decay into the lowest-lying excitonic states, and quasi-resonant (p-shell, etc.: process (b)) excitation, again accompanied by fast, non-radiative decay into the lowest-lying states.

2.2 Quantum Dot Electron Spin Qubits: Direct Manipulation

A single electron spin in a InAs quantum dot can act as a spin-qubit [15–17]. The introduction of a magnetic field $\vec{B}$ creates a natural quantization axis, independent of the growth axis, and will split the electron spin states due to the Zeeman energy:

$$H_{spin} = \frac{g}{2}\mu_B \vec{B} \cdot \vec{\sigma} \qquad (2.13)$$

where μ_B is the Bohr magneton, and g the Landé g-factor.[2] The total magnetic field experienced by the single quantum dot spin, can consist both of an externally applied field, and a net resulting field from the hyperfine interaction with the nuclear spins inside the quantum dot [2, 19–21] – the latter will be discussed in more detail in Chaps. 3 and 6.

An external, static magnetic field, strong enough to lift the degeneracy of the electron spin states and to overcome the randomly oriented hyperfine field, provides one axis of control for the qubit. The free evolution under the spin Hamiltonian, U can be written as a time-dependent rotation around an axis, $\vec{n}_{\vec{B}}$ that is aligned with the external magnetic field:

$$U = e^{-\frac{iH_{spin}t}{\hbar}} = e^{-\frac{ig\mu_B Bt}{\hbar}\left(\frac{\vec{n}_{\vec{B}} \cdot \vec{\sigma}}{2}\right)} \qquad (2.14)$$

In comparing Eq. 2.14 with Eq. 1.2, we see that the magnetic field does indeed give rise to a coherent rotation, with an angle that evolves linearly in time, and is proportional to the magnetic field strength: the well-known Larmor precession.

The static magnetic field can act as quantization axis, also providing one axis of coherent rotation control, yet another axis is necessary in order to have full SU(2)-control [3]. Equation 2.14 suggests that additional, differently oriented magnetic

---

[2]Technically, this is a tensorial quantity. For electrons, with their s-type symmetry, this tensor reduces to a constant times the unit tensor; for holes, however, the tensor is very much anisotropic – see e.g. [18]

Fig. 2.3 RF-spin control, and the rotating wave equation: in (**a**), a pulsed magnetic field is added to the DC field; due to the large DC field, only a small-angle precession is possible for initial spin-alignment along the DC field. In (**b**), the pulsed field is synchronized to (almost) coincide with the static Larmor precession. The net effect is a much larger angle precession

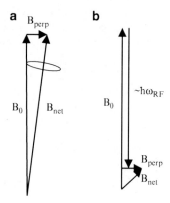

fields could do this. In fact, except for the magnetic field itself, no other symmetry-breaking elements are present in the spin-Hamiltonian: therefore, a fully vectorial, arbitrarily orientable magnetic field would provide full spin control. Unfortunately, switching on and off large magnetic fields is a slow and tedious process. Instead, small RF magnetic fields are typically added to the static field to realize full spin-control [3].

The advantage of RF-fields over simple pulsed ones is indicated in Fig. 2.3. For a small, off-axis, pulsed field, the effect of the large, static field limits any precession to a net axis, quasi-aligned with the DC field. However, by repeating the pulsation, and synchronizing it to the Larmor precession, a much larger net effect can be obtained. Denoting the magnetic field axis by x, combining $g\mu_B B$ as $\hbar\omega_L$, and with $\vec{B}_{RF} = B_{RF}[cos(\omega_{RF}t + \phi)\vec{y} + sin(\omega_{RF}t + \phi)\vec{z}]$,[3] we can re-write the spin Hamiltonian as follows:

$$H_{spin} = \frac{\hbar\Omega_L}{2}\sigma_x + \frac{\hbar\Omega_{Rabi}}{2}[cos(\omega_{RF}t + \phi)\sigma_y + sin(\omega_{RF}t + \phi)\sigma_z] \quad (2.15)$$

where we combined $g\mu_B B_{RF}$ as $\hbar\Omega_{Rabi}$. Transforming to a rotating frame, with the rotation axis along x and rotating at ω_{RF} yields:

$$H_{spin,rot} = \frac{\hbar(\Omega_L - \omega_{RF})}{2}\sigma_x + \frac{\hbar\Omega_{Rabi}}{2}[cos(\phi)\sigma_y + sin(\phi)\sigma_z] \quad (2.16)$$

From Eq. 2.16, we see that for resonant RF pulsation ($\omega_{RF} = \Omega_L$) the net effect (in the rotating frame) consists of a rotation around an axis perpendicular to the static magnetic field, where the exact rotation axis can be chosen by adjusting the

---

[3] While we choose a rotating magnetic field for convenience, any uniaxial field could be decomposed into two counter-rotating ones, after which a rotating wave approximation can be invoked to neglect one of them.

phase ϕ of the RF-field. Such direct, RF-magnetic field manipulation was realized in gate-defined quantum dots [22]. It is, however, still somewhat slow: for RF-frequencies of several GHz, at least several cycles are needed in order to flip the spin, leading to (far) sub-GHz spin manipulation speeds, in practice limited by the strength of the RF field. Higher speeds were obtained using clever combinations of hyperfine interactions and spin-orbit coupling, allowing direct electrical control [23]. However, there as well, several RF-cycles are needed to perform a spin-flip operation.

2.3 Quantum Dot Electron Spins: From Optical Dipole Interactions to Spin Control

One way to circumvent the Larmor-precession limitation to coherent control speed, is to use auxiliary states in the optical domain. Optical frequencies are in the 100s of THz range, and allow modulation on picosecond to femtosecond timescales. Self-assembled InAs quantum dots lend themselves particularly to this task, due to the presence of excited, optically active states. The level structure of a singly charged InAs quantum dot is indicated in Fig. 2.4, for the case of a single electron and a magnetic field perpendicular to the growth axis (Voigt geometry, oriented along x, Fig. 2.4b).

The relevant optical transition is indicated in Fig. 2.4a. The ground state consists of a single electron, spin-split by the magnetic field. After (virtual) absorption of a single photon, an excited state is created consisting of an electron and an exciton – a three-particle system (trion) consisting of two electrons and one hole [18]. Due to the strong confinement in the quantum dot, the electrons form a spin singlet, with the triplet states split by 10s of meV. As a singlet is magnetically inert, the magnetic properties of the trion are determined by those of the unpaired hole. This hole has heavy hole character, in view of the splitting off of the light holes because of confinement effects.

In the Voigt geometry, the Luttinger-Kohn Hamiltonian [5] mixes the heavy hole states, resulting in a small splitting characterized by an in-plane g-factor [18]. This is unlike the Faraday geometry, where field and quantization axis are aligned, leading to simple energy-renormalization and degeneracy lifting upon application of a strong magnetic field. The new eigenstates (spin part) can now be described as follows:

$$\text{electrons}: \frac{1}{\sqrt{2}}(|\uparrow\rangle_z \pm |\downarrow\rangle_z) = |\uparrow\rangle_x (|\downarrow\rangle_x) \qquad (2.17)$$

$$\text{trions}: \frac{1}{\sqrt{2}}(|\uparrow\downarrow\Uparrow\rangle_z \pm |\uparrow\downarrow\Downarrow\rangle_z) = |\uparrow\downarrow\Uparrow\rangle_x (|\uparrow\downarrow\Downarrow\rangle_x) \qquad (2.18)$$

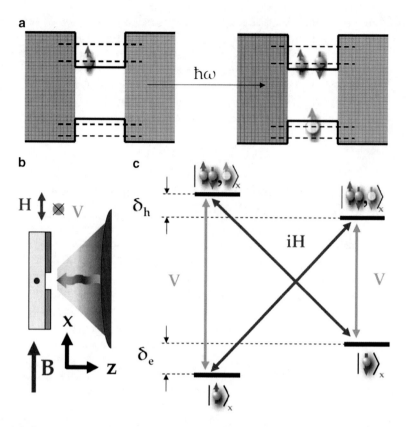

Fig. 2.4 Schematic overview of the level structure of a single-electron-charged quantum dot. (**a**) A single electron is resonantly excited into a trion state: one electron singlet (*blue arrows*) and one unpaired hole (*orange*). (**b**) Outline of the Voigt geometry used in this work: the magnetic field is aligned along the x-direction, perpendicular to the growth axis (z); H and V are in the growth plane, respectively parallel and perpendicular to the magnetic field orientation. (**c**) Level structure upon application of a Voigt geometry magnetic field; two Λ-systems emerge, which can be used to manipulate the spin state

Note that, in Eq. 2.18 and in the remainder of this work, $|\uparrow\downarrow\Uparrow\rangle_x$ ($|\uparrow\downarrow\Downarrow\rangle_x$) refer explicitly to trions consisting of a superposition of growth axis-aligned heavy holes: this superposition is *not* equivalent to a heavy hole oriented along the magnetic field. This difference is crucial with regards to the optical selection rules.[4]

---

[4]The description in terms of superpositions of pure heavy-hole eigenstates assumes that the angular momentum along the growth direction, J_z, is a good quantum number. In other words, cylindrical symmetry is assumed to be preserved. In view of the differences in confinement distances, this assumption is often but not always justified: strain and large asymmetries in particular quantum dots can lead to inmixing of light holes, which in turn affects the optical selection rules. We refer to Ref. [18] and references therein for more details. For the quantum dots used in the remainder of this work, pre-screening and selection of dots with 'clean' selection rules was applied.

2.3 Quantum Dot Electron Spins: From Optical Dipole Interactions to Spin Control

The selection rules for optical dipole transitions can be intuitively and semi-classically derived as follows in terms of allowed photonic emission processes: in view of the conservation of angular momentum along the growth direction, a trion state with the heavy hole spin pointed upward ($|\uparrow\downarrow\Uparrow\rangle_z$) will decay to an electron state with spin-up ($|\uparrow\rangle_z$), with the difference in angular momentum taken up by a circularly polarized ($\vec{\sigma}_+$) photon. Likewise, a trion with heavy hole pointed downward can couple to a spin-down electron state, emitting a left-hand circularly polarized photon. The other combinations would require a net change of two in angular momentum, and are forbidden – as long as higher-order effects such as heavy-light hole mixing can be ignored [18]. While a magnetic field in Voigt geometry breaks the cylindrical symmetry, its effect is typically much smaller than that of the large confinement/quantization, and can therefore be treated perturbatively – this is the approach described above, resulting in eigenstates that are coherent superpositions of the growth-direction defined states. We therefore have:

$$|\uparrow\downarrow\Uparrow\rangle_z \rightarrow |\uparrow\rangle_z \otimes \vec{\sigma}_+ \tag{2.19}$$

$$|\uparrow\downarrow\Downarrow\rangle_z \rightarrow |\downarrow\rangle_z \otimes \vec{\sigma}_- \tag{2.20}$$

$$|\uparrow\downarrow\Uparrow\rangle_x = \frac{1}{\sqrt{2}}(|\uparrow\downarrow\Uparrow\rangle_z + |\uparrow\downarrow\Downarrow\rangle_z) \rightarrow \frac{1}{\sqrt{2}}(|\uparrow\rangle_z \otimes \vec{\sigma}_+ + |\downarrow\rangle_z \otimes \vec{\sigma}_-) \tag{2.21}$$

$$|\uparrow\downarrow\Downarrow\rangle_x = \frac{1}{\sqrt{2}}(|\uparrow\downarrow\Uparrow\rangle_z - |\uparrow\downarrow\Downarrow\rangle_z) \rightarrow \frac{1}{\sqrt{2}}(|\uparrow\rangle_z \otimes \vec{\sigma}_+ - |\downarrow\rangle_z \otimes \vec{\sigma}_-) \tag{2.22}$$

While the first two equations describe the possible optical channels in Faraday geometry (Eqs. 2.19 and 2.20), the latter two (Voigt geometry) are technically entangled states between the spin and the photon, indicating that, upon spontaneous decay from the excited state, the interference between different pathways gives rise to spin-photon entanglement. We shall discuss this in more detail in Chap. 7. For now, it suffices to project the entangled state onto the Voigt-geometry, electron spin eigenstates:

$$|\uparrow\downarrow\Uparrow\rangle_x \rightarrow \frac{1}{\sqrt{8}}(\vec{\sigma}_+ + \vec{\sigma}_-) \otimes (|\uparrow\rangle_z + |\downarrow\rangle_z)$$

$$\rightarrow \frac{1}{2}[\frac{1}{\sqrt{2}}(\vec{Y} - i\vec{X}) + \frac{1}{\sqrt{2}}(\vec{Y} + i\vec{X})] \otimes |\uparrow\rangle_x$$

$$\rightarrow \frac{1}{\sqrt{2}}\vec{Y} \otimes |\uparrow\rangle_x \tag{2.23}$$

$$|\uparrow\downarrow\Uparrow\rangle_x \rightarrow \frac{1}{\sqrt{2}}i\vec{X} \otimes |\downarrow\rangle_x \tag{2.24}$$

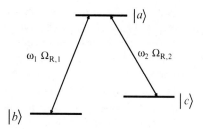

Fig. 2.5 Level structure used for coherent state manipulation through a Λ-system. $|c\rangle$ is the auxiliary, optically active state used for coherent control of the ground states $|a\rangle$ and $|b\rangle$. $\Omega_{1,2}$ indicate the Rabi frequency of the respective transition, proportional to the optical dipole elements ($\Omega = \mu \times E/\hbar$)

$$|\uparrow\downarrow\Downarrow\rangle_x \rightarrow \frac{1}{\sqrt{2}}\vec{Y} \otimes |\downarrow\rangle_x \qquad (2.25)$$

$$|\uparrow\downarrow\Downarrow\rangle_x \rightarrow \frac{1}{\sqrt{2}}i\vec{X} \otimes |\downarrow\rangle_x \qquad (2.26)$$

As we can see, each of the two trion states is connected to both of the electron spin ground states – this is the well-known Λ-configuration [24]. Also, the scaling factor of $\frac{1}{\sqrt{2}}$ indicates that the individual matrix elements in the Voigt geometry are reduced over those in Faraday geometry.[5] The selection rules in Voigt geometry are schematically illustrated in Fig. 2.4c. It is the Λ-configuration in Voigt geometry that allows for optical control of the spin state.

2.3.1 Coherent Spin Control: CPT and STIRAP

Figure 2.5 indicates a simplified, three-level Λ-system that can be coherently manipulated using optical dipole transitions. The ground states, $|b\rangle$ and $|c\rangle$ are connected to a single excited state, $|a\rangle$ through optical dipole elements with Rabi-frequencies $\Omega_{1,2}$ (the Rabi frequency is related to the dipole element μ: $\hbar\Omega = \mu \times E$). Following the analysis in [24], we can write the Hamiltonian as follows:

$$H = H_0 + H_1 \qquad (2.27)$$

---

[5] Given that the decay *probabilities* are proportional to the matrix elements squared (one can e.g. invoke a Fermi's golden rule argument [24]), this is actually a statement about the conservation of decay probability: for each excited state, in Voigt, there are two pathways for decay, each with half the probability of the single decay pathway in Faraday. These pathways do interfere, as we will show in Chap. 7.

2.3 Quantum Dot Electron Spins: From Optical Dipole Interactions to Spin Control

$$H_0 = \begin{pmatrix} \hbar\omega_a & 0 & 0 \\ 0 & \hbar\omega_c & 0 \\ 0 & 0 & \hbar\omega_c \end{pmatrix} \tag{2.28}$$

$$H_1 = \begin{pmatrix} 0 & -\frac{\hbar}{2}\Omega_1 e^{-i\omega_1 t} & -\frac{\hbar}{2}\Omega_2 e^{-i\omega_2 t} \\ -\frac{\hbar}{2}\Omega_1^* e^{i\omega_1 t} & 0 & 0 \\ -\frac{\hbar}{2}\Omega_2^* e^{i\omega_2 t} & 0 & 0 \end{pmatrix} \tag{2.29}$$

Working in the interaction picture, and with $\omega_1 = \omega_a - \omega_b$, $\omega_2 = \omega_a - \omega_c$, we obtain the following set of equations:

$$|\Psi\rangle = c_a(t)e^{-i\omega_a t}|a\rangle + c_b(t)e^{-i\omega_b t}|b\rangle + c_c(t)e^{-i\omega_c t}|c\rangle \tag{2.30}$$

$$i\hbar\dot{c} = H_{int}c$$

$$c = \begin{pmatrix} c_a(t) \\ c_b(t) \\ c_c(t) \end{pmatrix} \tag{2.31}$$

$$H_{int} = \begin{pmatrix} 0 & -\frac{\hbar}{2}\Omega_1 & -\frac{\hbar}{2}\Omega_2 \\ -\frac{\hbar}{2}\Omega_1^* & 0 & 0 \\ -\frac{\hbar}{2}\Omega_1^* & 0 & 0 \end{pmatrix} \tag{2.32}$$

One stationary solution of Eq. 2.31 has the following form:

$$c_{dark} = \begin{pmatrix} 0 \\ \frac{\Omega_2(t)}{\sqrt{|\Omega_1(t)|^2 + |\Omega_2(t)|^2}} \\ \frac{-\Omega_1(t)}{\sqrt{|\Omega_1(t)|^2 + |\Omega_2(t)|^2}} \end{pmatrix} \tag{2.33}$$

where we explicitly take any time-dependence of $\Omega_{1,2}$ into account.[6] We immediately notice that this state does not have any component of the excited, optically active state – hence its being referred to as a 'dark' state. In other words: if the system ends up in this dark superposition of the ground states $|b\rangle$ and $|c\rangle$, then it will always remain there, barring any incoherent effects involving the ground states. This effect is known as coherent population trapping (CPT), and has been observed in several systems, including electron-charged InAs quantum dots [26–28].

In addition, and for sufficiently *slow* (adiabatic) evolution of the optical control fields $\Omega_{1,2}$, the system will remain in its eigenstates by virtue of the adiabatic theorem of quantum mechanics [25]. As the dark state solution encompasses both states $|b\rangle$ ($\Omega_1 = 0$) and $|c\rangle$ ($\Omega_2 = 0$), adiabatically switching on and off of the control

---

[6]This is, strictly speaking, only valid for slow, adiabatic evolution of the Rabi frequencies $\Omega_{1,2}$ – we refer to [24, 25] for the necessary caveats.

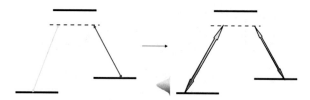

Fig. 2.6 From stimulated Raman transitions to ultrafast pulses: the two-laser scheme used for stimulated Raman control of the ground states can be replaced by a single, ultrafast and broadband laser-pulse scheme, where the polarization and bandwidth of the pulse are chosen so as to overlap with both dipole transitions

laser fields will coherently transfer the population from $|b\rangle$ to $|c\rangle$ or vice versa. This coherent transfer technique is extensively used in the atomic physics community, where it is known as Stimulated Raman Adiabatic Passage (STIRAP) [28].

2.3.2 Ultrafast Optical Control of Λ-Systems

While adiabatic passage in Λ-systems is a commonly established technique these days among the quantum optics community at large, it was recently realized that a variant hereof can be implemented using ultrafast, broadband pulses [15–17, 29–32]. We shall extensively elaborate on such ultrafast control in Chap. 3, where we shall show that full SU(2) control can be obtained using these techniques.

The basic idea is indicated in Fig. 2.6. CPT and STIRAP require two coherent light sources, carefully matched in frequency so as to account for the energy difference between the ground states, and often detuned from the excited states in order to ensure adiabaticity. Using one ultrafast pulse with a bandwidth exceeding the energy separation of the ground states, and with polarization chosen such that it encompasses both branches of the Λ-system, one can realize the stimulated Raman transitions as well. One semi-intuitive way to understand this, is to view the broadband pulse as a superposition of different single-frequency pairs, each separated by the ground-state energy difference. Each of these pairs then realizes a stimulated Raman transition with a certain amplitude and phase (the latter dependent on the arrival time of the pulse [29]), with the net effect of the pulse determined by the coherent superposition of the partial waves. Using such ultrafast pulses has resulted in all-optical spin control on picoseconds timescales [15, 16, 31, 32].

References

1. A. Imamoğlu et al. Quantum information processing using quantum dot spins and cavity QED. *Phys. Rev. Lett.*, 83:4204, 1999.
2. R. Hanson, L. P. Kouwenhoven, J. R. Petta, S. Tarucha, and L. M. K. Vandersypen. Spins in few electron quantum dots. *Rev. Mod. Phys.*, 79:1217, 2007.

References

3. M. A. Nielsen and I. L. Chuang. *Quantum Computation and Quantum Information*. Cambridge University Press, 2000.
4. D. Loss and D. P. DiVincenzo. Quantum computation with quantum dots. *Phys. Rev. A*, 57:120, 1998.
5. P. Yu and M. Cardona. *Fundamentals of Semiconductors - Physics and Materials Properties (3rd Edition)*. Springer, 2001.
6. V. N. Golovach, A. Khaetskii, and D. Loss. Phonon-Induced Decay of the Electron Spin in Quantum Dots. *Phys. Rev. Lett.*, 93:016601, 2004.
7. J. M. Elzerman, R. Hanson, L. H. Willems van Beveren, B. Witkamp, L. M. K. Vandersypen, and L. P. Kouwenhoven. Single-shot read-out of an individual electron spin in a quantum dot. *Nature*, 430:431, 2004.
8. M. Kroutvar, Y. Ducommun, D. Heiss, M. Bichler, D. Schuh, G. Abstreiter, and J. J. Finley. Optically programmable electron spin memory using semiconductor quantum dots. *Nature*, 432:81, 2004.
9. E. O. Kane. Band structure of Indium Antimonide. *Journal of Physics and Chemistry of Solids*, 1:249, 1957.
10. C. Santori, M. Pelton, G. Solomon, Y. Dale, and Y. Yamamoto. Triggered single photons from a quantum dot. *Phys. Rev. Lett.*, 86:1502, 2001.
11. C. Santori, D. Fattal, J. Vuckovic, G. S. Solomon, and Y. Yamamoto. Indistinguishable photons from a single-photon device. *Nature*, 419:594, 2002.
12. P. Michler, A. Kiraz, C. Becher, W. V. Schoenfeld, P. M. Petroff, L. Zhang, E. Hu, and A. Imamoglu. A quantum dot single-photon turnstile device. *Science*, 290:2282, 2000.
13. M. Pelton, C. Santori, J. Vuckovic, B. Zhang, G. S. Solomon, J. Plant, and Y. Yamamoto. Efficient source of single photons: A single quantum dot in a micropost microcavity. *Phys. Rev. Lett.*, 89:233602, 2002.
14. E. Moreau, I. Robert, L. Manin, V. Thierry-Mieg, J. M. Gérard, and I. Abram. A single-mode solid-state source of single photons based on isolated quantum dots in a micropillar. *Physica E*, 13:418, 2002.
15. J. Berezovsky, M. H. Mikkelsen, N. G. Stoltz, L. A. Coldren, and D. D. Awschalom. Picosecond coherent optical manipulation of a single electron spin in a quantum dot. *Science*, 320:349, 2008.
16. D. Press, T. D. Ladd, B. Zhang, and Y. Yamamoto. Complete quantum control of a single quantum dot spin using ultrafast optical pulses. *Nature*, 456:218, 2008.
17. D. Press, K. De Greve, P. McMahon, T. D. Ladd, B. Friess, C. Schneider, M. Kamp, S. Höfling, A. Forchel, and Y. Yamamoto. Ultrafast optical spin echo in a single quantum dot. *Nat. Photonics*, 4:367, 2010.
18. M. Bayer et al. Fine structure of neutral and charged excitons in self-assembled In(Ga)As/(Al)GaAs quantum dots. *Phys. Rev. B*, 65:195315, 2002.
19. A. Greilich et al. Nuclei-induced frequency focusing of electron spin coherence. *Science*, 317(4):1896, 2007.
20. W. A. Coish and D. Loss. Hyperfine interaction in a quantum dot: Non-Markovian electron spin dynamics. *Phys. Rev. B*, 70:195340, 2004.
21. W. M. Witzel and S. Das Sarma. Quantum theory for electron spin decoherence induced by nuclear spin dynamics in semiconductor quantum computer architectures: Spectral diffusion of localized electron spins in the nuclear solid-state environment. *Phys. Rev. B*, 74:035322, 2006.
22. F. H. L. Koppens, C. Buizert, K. J. Tielrooij, I. T. Vink, K. C. Nowack, T. Meunier, L. P. Kouwenhoven, and L. M. K. Vandersypen. Driven coherent oscillations of a single electron spin in a quantum dot. *Nature*, 442, 766.
23. K. C. Nowack, F. H. L. Koppens, Yu. V. Nazarov, and L. M. K. Vandersypen. Coherent Control of a Single Electron Spin with Electric Fields. *Science*, 318:1430, 2007.
24. M. O. Scully and M. S. Zubairy. *Quantum optics*. Cambridge University Press, 1997.
25. A. Messiah. *Quantum mechanics*. Dover, 1999.

26. K.-M. C. Fu, C. Santori, C. Stanley, M. C. Holland, and Y. Yamamoto. Coherent Population Trapping of Electron Spins in a High-Purity n-Type GaAs Semiconductor. *Phys. Rev. Lett.*, 95:187405, 2005.
27. X. Xu *et al*. Optically controlled locking of the nuclear field via coherent dark-state spectroscopy. *Nature*, 459(4):1105, 2009.
28. M. Fleischauer, A. Imamoglu, and J. P. Marangos. Electromagnetically induced transparency: Optics in coherent media. *Rev. Mod. Phys.*, 77:633, 2005.
29. S. M. Clark, K-M. C. Fu, T. D. Ladd, and Y. Yamamoto. Quantum computers based on electron spins controlled by ultrafast off-resonant single optical pulses. *Phys. Rev. Lett.*, 99:040501, 2007.
30. C. E. Pryor and M. E. Flatté. Predicted ultrafast single-qubit operations in semiconductor quantum dots. *Appl. Phys. Lett.*, 88:233108, 2006.
31. K. De Greve, P. L. McMahon, D. Press, T. D. Ladd, D. Bisping, C. Schneider, M. Kamp, L. Worschech, S. Höfling, A. Forchel, and Y. Yamamoto. Ultrafast coherent control and suppressed nuclear feedback of a single quantum dot hole qubit. *Nat. Phys.*, 7:872, 2011.
32. K.-M. C. Fu *et al*. Ultrafast control of donor-bound electron spins with single detuned optical pulses. *Nat. Phys.*, 4:780, 2008.

Chapter 3
Ultrafast Coherent Control of Individual Electron Spin Qubits

In this chapter, we shall derive how ultrafast optical control can be used in combination with Larmor precession for full SU(2) control of a single quantum dot electron spin. Such ultrafast electron spin control was first reported in Refs. [1–3]. In combination with accurate timing control over the optical fields used to realize this SU(2) control, arbitrary pulse (control) patterns can be applied to the spin [3]. As we will show in Sect. 3.3.2, such pulse sequences can be used to overcome the effects of slowly varying Larmor precession due to variations in the spin's solid state environment. As a particular example, we will study the effects of the hyperfine interaction between an electron spin and the nuclear spins inside the quantum dot [4–7] – the latter will be discussed in detail in Sect. 3.3.1.

3.1 Ultrafast Control: Operation Principle

3.1.1 Stimulated-Raman Picture

While CPT and STIRAP require two different laser sources in order to manipulate a Λ-system, ultrafast control is based on the notion that one single optical pulse, with bandwidth larger than the separation of the two ground states δ_e, contains a semi-continuum of single-frequency pairs that each fulfill the resonance condition $\omega_1 - \omega_2 = \delta_e$ (ω refers here to the frequency of the respective single-frequency pairs). Under appropriate conditions, all these pairs can be shown to constructively interfere [8]. The level structure and optical Rabi frequencies for a generalized ultrafast control scheme of an optical Λ-system are indicated in Fig. 3.1. δ_e refers to the energy separation of the ground state, Δ to the detuning of the laser frequency, and $\Omega_{1,2}$ are the optical Rabi frequencies associated with the optical dipole elements, derived from the interaction of a laser source at frequency ω_1 with the Λ-system.

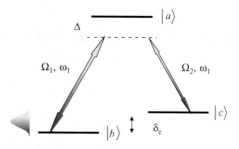

Fig. 3.1 Generalized level structure for the simplest possible stimulated Raman transition using one, broadband pulse. $|a\rangle$, $|b\rangle$ and $|c\rangle$ indicate the levels used, in Λ-configuration. $\hbar\delta_e$ is the energy splitting of the ground states, while $\hbar\Delta$ stands for the detuning of the optical field used. The field frequency is ω_1, and its matrix elements are Ω_1 and Ω_2

The system is described by the following Hamiltonian:

$$i\hbar \frac{d}{dt}|\Psi(t)\rangle = H|\Psi(t)\rangle$$

$$H = \begin{pmatrix} \hbar\omega_a & -\frac{\hbar}{2}\Omega_1 e^{-i\omega_1 t} & -\frac{\hbar}{2}\Omega_2 e^{-i\omega_2 t} \\ -\frac{\hbar}{2}\Omega_1^* e^{i\omega_1 t} & \hbar\omega_b & 0 \\ -\frac{\hbar}{2}\Omega_2^* e^{i\omega_2 t} & 0 & \hbar\omega_c \end{pmatrix} \quad (3.1)$$

Moving to the interaction picture, and invoking a rotating wave approximation, the equation of motion takes the following form:

$$|\Psi(t)\rangle = a(t)|\tilde{a}\rangle + b(t)|\tilde{b}\rangle + c(t)|\tilde{c}\rangle$$

$$i\hbar \begin{pmatrix} \dot{a}(t) \\ \dot{b}(t) \\ \dot{c}(t) \end{pmatrix} = \begin{pmatrix} \hbar\Delta & -\frac{\hbar}{2}\Omega_1(t) & -\frac{\hbar}{2}\Omega_2(t) \\ -\frac{\hbar}{2}\Omega_1^*(t) & \hbar\delta_e & 0 \\ -\frac{\hbar}{2}\Omega_2^*(t) & 0 & 0 \end{pmatrix} \begin{pmatrix} a(t) \\ b(t) \\ c(t) \end{pmatrix} \quad (3.2)$$

In Eq. 3.2, we can adiabatically eliminate the excited state ($\dot{a}(t) = 0$), provided that the detuning is much larger than the Rabi-frequencies $\Omega_{1,2}$:

$$i\hbar\dot{a}(t) = \hbar\Delta a(t) - \frac{\hbar}{2}\Omega_1(t)b(t) - \frac{\hbar}{2}\Omega_2 c(t) = 0$$

$$\rightarrow a(t) = \frac{\Omega_1(t)}{2\Delta}b(t) + \frac{\Omega_2(t)}{2\Delta}c(t). \quad (3.3)$$

Incorporating this back into Eq. 3.2, we obtain the following equation of motion for the ground states:

$$i\begin{pmatrix} \dot{b}(t) \\ \dot{c}(t) \end{pmatrix} = \begin{pmatrix} \delta_e - \frac{|\Omega_1|^2}{4\Delta} & -\frac{\Omega_1^*\Omega_2}{4\Delta} \\ -\frac{\Omega_2^*\Omega_1}{4\Delta} & -\frac{|\Omega_2|^2}{4\Delta} \end{pmatrix} \begin{pmatrix} b(t) \\ c(t) \end{pmatrix}. \quad (3.4)$$

3.1 Ultrafast Control: Operation Principle

The effect of the optical interaction is two-fold: there is an energy-renormalization of the ground states due to the AC-Stark shift, as well as an effective coupling of the two ground states due to the detuned interaction with the excited state.

For $|\Omega_1(t)| = |\Omega_2(t)|$, the effect of the AC-Stark shift is an over-all shift of the energy reference (global phase factor developing), which we can normalize out.[1] Defining $\Omega_{\text{eff}} = \frac{\Omega_1^*(t)\Omega_2(t)}{2\Delta}$, we therefore obtain:

$$i \begin{pmatrix} \dot{b}(t) \\ \dot{c}(t) \end{pmatrix} = \begin{pmatrix} \delta_e & -\frac{\Omega_{\text{eff}}}{2} \\ -\frac{\Omega_{\text{eff}}^*}{2} & 0 \end{pmatrix} \begin{pmatrix} b(t) \\ c(t) \end{pmatrix}. \tag{3.5}$$

Moving to a frame rotating at the difference frequency between the ground states, δ_e, the following equation of motion can be found:

$$|\Psi_{\text{spin}}(t)\rangle = b_{\text{rot}}(t)e^{-i\delta_e t}|\tilde{b}\rangle + c_{\text{rot}}|\tilde{c}\rangle$$

$$i \begin{pmatrix} \dot{b}_{\text{rot}} \\ \dot{c}_{\text{rot}} \end{pmatrix} = \begin{pmatrix} 0 & -\frac{\Omega_{\text{eff}} e^{i\delta_e t}}{2} \\ -\frac{\Omega_{\text{eff}}^* e^{-i\delta_e t}}{2} & 0 \end{pmatrix} \begin{pmatrix} b_{\text{rot}} \\ c_{\text{rot}} \end{pmatrix} \tag{3.6}$$

Moving to a (pseudo-)spin picture, with the (pseudo-)magnetic field along the z-direction, we can derive the following effective Hamiltonian in our rotating frame (rotating at the Larmor-precession frequency):

$$H_{eff} = \hbar \begin{pmatrix} 0 & -\frac{\Omega_{\text{eff}} e^{i\delta_e t}}{2} \\ -\frac{\Omega_{\text{eff}}^* e^{-i\delta_e t}}{2} & 0 \end{pmatrix}$$

$$= \hbar \begin{pmatrix} 0 & -\frac{|\Omega_{\text{eff}}|}{2} e^{i(\phi+\delta_e t)} \\ -\frac{|\Omega_{\text{eff}}|}{2} e^{-i(\delta_e t+\phi)} & 0 \end{pmatrix} \tag{3.7}$$

$$= \frac{-\hbar|\Omega_{\text{eff}}|}{2} \Big[\cos(\phi+\delta_e t)\sigma_x - \sin(\phi+\delta_e t)\sigma_y\Big]. \tag{3.8}$$

The evolution operator in the rotation frame, $U_{\text{eff}} = e^{-iH_{\text{eff}}t/\hbar}$, now has the following form, for a pulse of duration τ that is short:

$$U_{\text{eff}} = e^{-i|\Omega_{\text{eff}}|\tau(\frac{\cos(\phi+\delta_e t)\sigma_x - \sin(\phi+\delta_e t)\sigma_y}{2})} \tag{3.9}$$

$$U_{\text{eff}} = e^{-i\theta(\frac{\vec{n}(t)\cdot\vec{\sigma}}{2})} = R_{\vec{n}(t)}(\theta) \tag{3.10}$$

$$\vec{n}(t) = (\cos(\phi+\delta_e t), -\sin(\theta+\delta_e t), 0) \tag{3.11}$$

$$\theta = |\Omega_{\text{eff}}|\tau \tag{3.12}$$

---

[1] For $|\Omega_1(t)| \neq |\Omega_2(t)|$, there is a net difference in the energy of both ground states, which adds to the energy splitting δ_e. In the case of a spin qubit, as we shall derive later, this corresponds to a net magnetic field being applied, parallel to the field causing the Zeeman splitting in the first place.

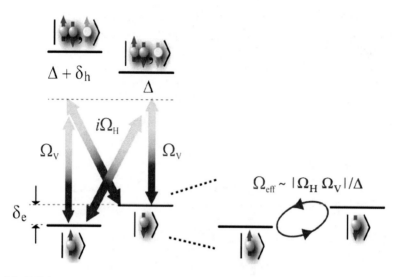

Fig. 3.2 Full four-level structure of an electron-charged quantum dot in Voigt geometry. Constructive interference of both Λ-systems results in a net coherent rotation of the electron spin, characterized by Ω_{eff} (Reproduced from [2] with permission, copyright Nature Publishing group (2008))

Here, we used the definition of a coherent spin rotation, Eq. 1.2. For a pulse that is sufficiently short compared to the Larmor precession frequency, the net effect, in the rotating frame, is a rotation over an angle $|\Omega_{\text{eff}}|\tau$, around an axis which depends on the arrival time of the pulse. Moving back to the lab frame, the pulse defines a rotation axis that is fixed in space, but the spin precesses at the Larmor precession frequency – the effect of the pulse therefore also depends on the arrival time.

3.1.1.1 From Three- to Four-Level Structure

In reality, the level-structure of an electron charged quantum dot in Voigt geometry has, to lowest order, four levels rather than three. There are two Λ-systems, one for each trion state, as indicated in Fig. 3.2. Moving to a frame where the magnetic field is along the x-direction (in plane), and the growth direction is defined as z (see also Fig. 2.4b), we can describe the electric field (polarization) as $E_H \vec{X} + E_V \vec{Y}$, where E_H and E_V are unit-length complex quantities, accounting for particular phase differences that reflect a generalized, elliptical polarization.

Denoting the respective trion states as $|\uparrow\downarrow\Uparrow\rangle_x$ and $|\uparrow\downarrow\Downarrow\rangle_x$, and with $\Omega_{H,V} = \frac{|\Omega|E_{H,V}}{\sqrt{2}}$ (reflecting the strength of the optical dipole elements in Voigt geometry), we can use Eqs. 2.23–2.26 to obtain the following result:

$$\Omega_{\uparrow\downarrow\Uparrow} = \frac{-i\Omega_H^*\Omega_V}{4\Delta + \delta_h} \simeq \frac{-i|\Omega|^2 E_H^* E_V}{4\Delta} \quad (3.13)$$

3.1 Ultrafast Control: Operation Principle

$$\Omega_{\uparrow\downarrow\downarrow} = \frac{i\Omega_H\Omega_V^*}{4\Delta} \cong \frac{i|\Omega|^2 E_H E_V^*}{4\Delta} \tag{3.14}$$

$$\Omega_{\text{eff}} = \Omega_{\uparrow\downarrow\uparrow} + \Omega_{\uparrow\downarrow\downarrow} = \frac{i|\Omega|^2}{4\Delta}(E_V^* E_H - E_H^* E_V) \tag{3.15}$$

From Eq. 3.15, we see that linearly polarized light cannot perform coherent rotations, and that circularly polarized light ($E_H = \pm i E_V$) maximizes the effective Rabi frequency:

$$\Omega_{\text{eff},\sigma^{\pm}} = \mp \frac{|\Omega|^2}{4\Delta} \tag{3.16}$$

The rotation angle θ, in turn, is approximately given by

$$\theta = \int_{pulse} \Omega_{\text{eff}} dt \tag{3.17}$$

$$\sim \int_{pulse} \frac{P(t)}{\Delta}$$

$$\sim \frac{\varepsilon_{pulse}}{\Delta} \tag{3.18}$$

where ε_{pulse} stands for the energy of a single pulse (this equation is valid so long as the pulse duration is short compared to the Larmor precession frequency, and in the limit where the conditions for adiabatic elimination are fullfilled: $\Delta \gg \Omega$). When changing the pulse energy, and working in our reference frame rotating at the Larmor precession frequency, one would therefore expect the spin to precess around a rotation axis determined by the arrival time of the pulse – Rabi-flopping, which shall be discussed in Sect. 3.2.

3.1.2 AC-Stark Shift Picture

A different, yet equivalent interpretation of the effect of an ultrafast pulse is based on the AC-Stark effect and the presence of a dark state when applying a strong, instantaneous light pulse with circular polarization. Working in the z-basis, along the growth direction, the level structure is indicated in Fig. 3.3. This is not an eigenbasis in the presence of the magnetic; the magnetic field along x will therefore create coherences between the z-basis states. Solving the Hamiltonian in the presence of the magnetic field, but in the absence of any pulse, would result in the Voigt-geometry eigenstates which we obtained before.

However, for an instantaneous, strong light pulse, applied much faster than the Larmor-period, we can ignore the dynamics due to the magnetic field. As one of the two z-basis spin states is dark for a circularly polarized light pulse, due to the optical selection rules, only the bright state is affected by the pulse. As the circularly

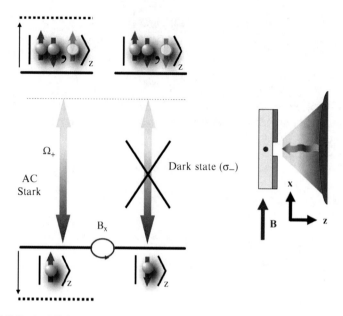

Fig. 3.3 AC-Stark shift interpretation of coherent pulse control. For circularly polarized light, there exists a dark state for the spin ground states of the quantum dot system. While the dark state remains unaffected, the orthogonal, bright state is AC-Stark shifted. Starting from an initial state which is a superposition of the dark state and the bright state, the resulting phase factor give rise to coherent spin oscillations. *Inset*: orientation of the magnetic field (Reproduced from [9])

polarized pulse couples only to a single trion state, the effect of the pulse on the two-level subsystem formed by $|\uparrow\rangle_z - |\uparrow\downarrow\Uparrow\rangle_z$ is the following (we assume again a detuning Δ):

$$|\Psi_{2-\text{level}}\rangle = a(t)|\uparrow\downarrow\tilde{\Uparrow}\rangle + b(t)|\tilde{\uparrow}\rangle$$

$$i\begin{pmatrix} \dot{a}(t) \\ \dot{b}(t) \end{pmatrix} = \begin{pmatrix} \Delta & -\frac{\Omega_+}{2} \\ -\frac{\Omega_+^*}{2} & 0 \end{pmatrix} \begin{pmatrix} a(t) \\ b(t) \end{pmatrix} \quad (3.19)$$

where we invoked the rotating wave approximation, and work in the interaction picture (reflected by the transformation $|\Uparrow\rangle \to |\uparrow\downarrow\Uparrow\rangle$). In view of the optical selection rules derived above, $|\Omega_+| = |\Omega|$. If the detuning, Δ, is much larger than the Rabi frequency, Ω_+, then the dominant effect of the pulse consists of an energy-renormalization of the spin and trion states known as the AC-Stark effect (to first order, no real excitation of the excited state). The AC-Stark shift of the electron spin state (downwards), $\delta_\uparrow$ is given by

$$\delta_\uparrow = \frac{\Delta}{2}\left(\sqrt{1 + \frac{|\Omega|^2}{\Delta^2}} - 1\right)$$

$$\simeq \frac{|\Omega|^2}{4\Delta}$$

$$\simeq \Omega_{\text{eff},\sigma^+}. \tag{3.20}$$

On the $|\uparrow\rangle_z, |\downarrow\rangle_z$-subsystem, this results in the following net interaction Hamiltonian:

$$H_{\text{eff}} = \begin{pmatrix} -\Omega_{\text{eff}} & 0 \\ 0 & 0 \end{pmatrix} \tag{3.21}$$

which we can renormalize by shifting the energy reference:

$$\frac{\tilde{H}_{\text{eff}}}{\hbar} = \begin{pmatrix} -\frac{\Omega_{\text{eff}}}{2} & 0 \\ 0 & \frac{\Omega_{\text{eff}}}{2} \end{pmatrix} \tag{3.22}$$

$$= -\frac{\Omega_{\text{eff}}}{2} \sigma_z \tag{3.23}$$

where z indicated the growth direction.

With the evolution operator, U_{eff} in our fixed labframe, we see that this again corresponds to an effective rotation:

$$U_{\text{eff}} = e^{-i \int_{pulse} H_{\text{eff}} dt / \hbar} \tag{3.24}$$

$$= e^{-i\theta(\frac{\sigma_z}{2})}$$

$$= R_z(\theta)$$

$$\theta = \int_{pulse} -\Omega_{\text{eff}} dt \simeq \frac{\varepsilon_{pulse}}{\Delta} \tag{3.25}$$

This is exactly the same result as was obtained using the Stimulated-Raman formalism. The AC-Stark interpretation clarifies the fact that, in the fixed, laboratory frame, the circularly polarized pulse realizes a coherent rotation around a fixed axis, corresponding to the growth and light propagation direction.

The effect of the strong light pulse can therefore be interpreted as the application of an instantaneous magnetic field in the z-direction, much large than the applied field in the x-direction that gives rise to the Zeeman splitting and Larmor precession.

3.2 All-Optical SU(2) Control

While the fast, broadband pulses can be used for generating coherent rotations around an axis along the z-direction, with a rotation angle determined by the pulse energy used, the magnetic field induces Larmor precession around the x-axis, with

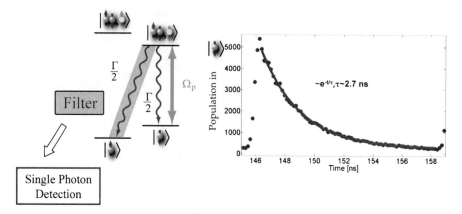

Fig. 3.4 Schematic illustration of the optical pumping scheme used for initialization and readout. A narrowband CW laser optically pumps the system into the $|\uparrow\rangle$-state, with a characteristic lifetime of about 2.7 ns. A single, linearly polarized photon along the $|\uparrow\downarrow\Downarrow\rangle - |\uparrow\rangle$-transition heralds the presence of population in the $|\downarrow\rangle$-state, and can be measured with about 0.1 % overall efficiency (Reproduced from [2] with permission, copyright Nature Publishing Group (2008))

the rotation angle proportional to the duration of the Larmor precession. As these two operations involve two orthogonal axes, with arbitrary angle control around both, any coherent spin control operation can be decomposed into a combination of them [2, 8, 10].

3.2.1 Initialization and Readout

Besides coherently controlling the qubit, one must also be able to initialize or reset the qubit efficiently, and read it out. Several possible readout schemes have been proposed and implemented, ranging from dispersive, non-demolition redout based on optical Kerr- or Faraday-rotation [11, 12] to resonance-fluorescence based read-out schemes [13]. While such techniques each have their merits, the Kerr- or Faraday-rotation suffers from low signal-to-noise ratios, and requires averaging over many experiments in order to obtain information over the spin state. Resonance-fluorescence on the other hand, only occurs efficiently in Faraday-geometry: in Voigt-geometry, the Λ-configuration will lead to optical pumping [14].

We use optical pumping for initializing the spin state [2, 14, 15]. The level structure and optical transitions used are indicated in Fig. 3.4. A narrowband, CW laser, resonant with the $|\uparrow\downarrow\Downarrow\rangle - |\downarrow\rangle$-transition, is used. Upon excitation of the $|\uparrow\downarrow\Downarrow\rangle$-state, spontaneous emission from the Λ-system allows for the photon to decay along two different branches (or superpositions hereof: see Chap. 7). When the decay occurs along the $|\uparrow\downarrow\Downarrow\rangle$-$|\downarrow\rangle$-branch, then the laser will re-excite the system into the $|\uparrow\downarrow\Downarrow\rangle$-state, until eventually a single linearly polarized photon is scattered off the

$|\uparrow\downarrow\Downarrow\rangle$-$|\uparrow\rangle$-branch: after this, the system is stuck in the $|\uparrow\rangle$-state, which is a dark state for the pumping process.

The time-constant for this optical pumping process reflects the fact that several cycles of emission into the 'wrong' branch are possible before the actual pumping into the dark state happens: for the experimental data in Fig. 3.4 (same quantum dot as was used for the experiments described in Chap. 7), the decay is exponential, with a decay time of 2.7 ns, compared to about 600 ps of spontaneous emission lifetime for the quantum dot used in that experiment. Regardless, optical pumping can initialize the spin with high fidelity, within a few nanoseconds, which is orders of magnitude faster than through spontaneous spin flips (T_1-process [16]), and much better (higher fidelity) than the spontaneous spin polarization that would be expected by virtue of the low temperature (1.6 K) and energetic splitting ($\sim$0.08 meV). We refer to Appendix A.1.1 for a detailed analysis of the fidelity of optical pumping.

Crucially, during the optical pumping process, a single photon is scattered with different frequency and opposite polarization compared to the $|\uparrow\downarrow\Downarrow\rangle$-$|\downarrow\rangle$-branch. By polarization and frequency filtering, this single photon can be distinguished from laser scatter, and can be used to herald the presence of a spin-down electron spin (system in the $|\downarrow\rangle$-state). However, due to detector inefficiencies, total internal reflection inside the sample, and imperfect transmission through the optical collection path, there is only a 0.1 % chance for the single photon to be detected by a single photon detector. Hence, while a detector click faithfully heralds the presence of a spin-down spin (modulo detector dark counts, which are generally negligible compared to the real count rates in our system), the probability of success is very low, and requires many cycles of the same experiment to be implemented before accurate state information can be obtained.

The combination of optical pumping for readout and initialization, and Larmor precession and ultrafast optical control for coherent spin manipulation, allows for all-optical manipulation of an electron-spin qubit [1–3] – we refer to Fig. 3.5 for an overview of all processes involved.

3.2.2 Device Design and Experimental Setup

The sample design used is indicated in Fig. 3.6, and was extensively described previously [3]. The samples are MBE grown, and contain on average about 2×10^9 cm^{-2} or fewer self-assembled quantum dots, about 30 nm in diameter, that are formed using the Straski-Krastanov method. The quantum dots are located at the center (antinode) of a planar GaAs microcavity. A δ-doping layer of about 1×10^{10} cm^{-2} silicon donors was grown 10 nm below the quantum dot layer: some of those donors are close enough to a quantum dot for the charge to get trapped inside a quantum dot. This *probabilistic* charging scheme can be verified by magnetophotoluminescence studies [2, 3], which can be used to prescreen for charged quantum dots. In practice, approximately one-third of the quantum dots

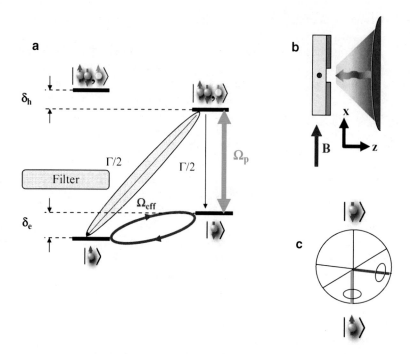

Fig. 3.5 All-optical control of an electron spin qubit. (**a**) Optical pumping realizes initialization and spin readout, while a broadband, ultrafast laserpulse induces an effective rotation around the z-axis (lab frame). In addition, an x-axis oriented magnetic field causes Larmor precession. (**b**) Schematic viewgraph of the orientational conventions used. (**c**) While the magnetic field realizes a coherent rotation around the x-axis (*green*), the optical pulse realizes coherent control around the z-axis (*red*). With arbitrary angle control, this corresponds to full SU(2) control

of an average wafer are charged. The lower and upper cavity mirrors contain 24 and 5 pairs of $Al_{0.9}Ga_{0.1}As$/GaAs $\lambda/4$ layers, respectively, acting as dielectric mirrors, and giving the cavity a quality factor of about 200. The cavity increases the signal-to-noise ratio of the measurement in two ways. First, it increases the collection efficiency by directing most of the quantum dot emission towards the objective lens (highly reflective back-mirror) and, second, it reduces the laser power required to achieve optical pumping, thereby reducing the reflected pump-laser noise. This planar microcavity is capped with a 100-nm-thick aluminium mask with 10 μm windows for optical access. These mask holes are significantly larger than the diffraction-limited spot size of the lasers used, and serve mainly as alignment markers on the sample. The selection of a single quantum dot occurs by virtue of the low quantum dot density within the diffraction-limit excitation and collection spot (some 10 quantum dots in the 1 μm spot size), in combination with their spectral

3.2 All-Optical SU(2) Control

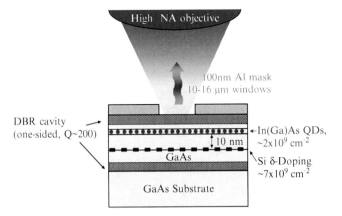

Fig. 3.6 Device design used for all-optical control of a single electron spin qubit. InAs quantum dots are embedded within an optical microcavity that preferentially redirects the spontaneously emitted light towards the collection optics. Silicon δ-doping stochastically dopes a single quantum dot, which is coherently controlled

inhomogeneity. As the cavity selects out a spectral region of some 5–10 nm only, there is typically only one quantum dot resonant with the cavity.[2]

The optical setup used for coherent and all-optical spin manipulation is indicated in Fig. 3.7, and was also described previously [3]. The sample is mounted in a superconducting magnetic cryostat (Oxford Spectromag, base temperature: 1.5 K) with a variable magnetic field up to 10 T. An aspheric objective lens with numerical aperture (NA) of 0.68 is mounted inside the cryostat to focus both pump and rotation lasers onto the sample, and collects the photoluminescence. Control over the sample position inside the cryostat (selection of a particular quantum dot) requires the use of piezo-electric 'slip-stick' positioners (Attocube systems). Single-photon photoluminescence is collected through the objective lens and directed onto a single-photon counter by means of a confocal microscopy setup (pinhole size: 50 μm) that rejects scattered light originating outside the diffraction-limited spot. The quantum dot photoluminescence emission is subsequently spectrally filtered using a home-built double-monochromator with a resolution of 0.02 nm, and detected on a commercial single-photon counter (SPCM, Perkin-Elmer/Excelitas). Scattered laser light is further rejected by means of cross-polarization. For the ultrafast

---

[2]The distribution of quantum dot emission wavelengths reflects their size- and strain distribution due to the self-assembled growth process. In practice, this distribution peaks around 880 nm, and tails off very slowly towards 950–1,000 nm. For quantum dots in the 940 nm range, as were used for the spin echo experiments described in Sect. 3.3.2, there is often less than one quantum dot resonant with the cavity. For the more blue-shifted dots used for spin-photon entanglement verification in Sect. 7, the overall dot density was reduced, yet the cavity is closer to the peak of the distribution. Therefore, typically, more than one quantum dot could be found resonant with the cavity. However, their spectral inhomogeneity still allows for selective excitation of and collection from one particular dot.

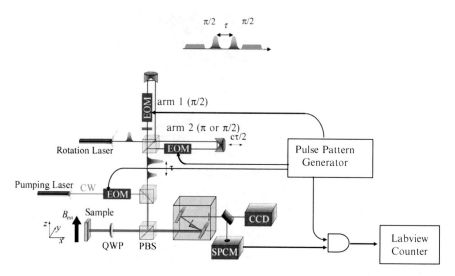

Fig. 3.7 The optical setup used for all-optical electron spin control. See text for details. Acronyms used: *QWP* quarter-waveplate, *PBS* polarizing beamsplitter, *CCD* charge-coupled device, *SPCM* single photon counting module (Figure reproduced from [3]). *Inset*: control sequence performed using the setup

coherent control, a modelocked laser (Spectra-Physics Tsunami) is used, which outputs 3-ps pulses every 13.2 ns (75.76 MHz repetition rate). The wavelength can be set arbitrarily, and is typically red-detuned by about 1 nm from the trion-transitions. The rotation laser path is divided into multiple arms, each of which can be arbitrarily controlled in intensity (pulse rotation angle) and delay (effective rotation axis in the rotating basis of the spin: see Sect. 3.1.1). A pair of free-space electro-optic modulators (EOMs, Conoptics) are used as in order to pulse-pick from the 75.76 MHz pulse train in each branch. Each EOM is double-passed to achieve extinction ratios of 10^4 or higher, while computer-controlled stages allow fine control over the arrival time of each pulse (the rotation laser acts as the master oscillator). The optical pumping laser is gated by a fiber-based EOM (EOSpace) with an extinction ratio of some 10^4; the optical pumping time used varies between 13 ns (single cycle of the master-oscillator laser) or 26 ns (two cycles). All EOMs are controlled by a data pattern generator (Tektronix) synchronized to the modelocked rotation laser. The rotation laser can be chopped at 1 kHz by electronically gating the EOMs, and this modulation can then be used for a digital lock-in procedure, by checking for a detector count-signal synchronized to this frequency. To reject detector dark counts, the photon counter is gated on only during the optical pumping (readout) pulse.

3.2 All-Optical SU(2) Control

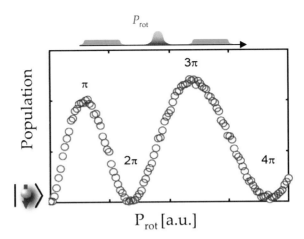

Fig. 3.8 Rabi-oscillations of a single electron spin qubit, indicating arbitrary-angle control of a rotation around the z-axis (lab frame). *Inset*: pulse sequence used

3.2.3 Experimental Results

3.2.3.1 Rabi-Oscillations

As we discussed before, a single optical pulse generates a coherent spin rotation around a fixed axis in the lab frame (which, in the frame rotating with the spin at the Larmor precession, comes down to an axis which varies depending on the arrival time of the pulse – see Sect. 3.1.1). The rotation pulse angle, θ, in the limit of short pulse durations compared to the Larmor-precession frequency, is proportional to $\varepsilon_{pulse}/\Delta$, where ε_{pulse} stands for the pulse energy used. Hence, as a function of varying rotation pulse power, one would expect to see coherent oscillations in the spin state, reflecting the continuously increasing rotation angle: the famous Rabi-oscillations of a single spin [1–3, 9, 15].

Figure 3.8 illustrates such Rabi-oscillations. The rotation laser used has a pulse duration of about 3 ps, detuned by about 1 nm from the trion-transitions, and its power (energy) is continuously increased. After initialization into the $|\uparrow\rangle$-state by optical pumping, the population in $|\downarrow\rangle$ is measured, as a function of the rotation pulse power used. The signal is indeed oscillatory, and the slight deviation from perfect sinusoidal behavior can be explained in terms of the non-linearity of the AC-Stark shift (Ω_{eff} is a slightly non-linear function of rotation pulse energy, see Eq. 3.20) and deviations from the 'ultrafast' approximation (competition between the concurrent Larmor precession frequency and the effect of the pulse-induced rotation for non-instantaneous pulses). We refer to Refs. [2] and [9] for a more detailed analysis.

3.2.3.2 Ramsey-Fringes

While the optical control pulses generate an arbitrary-angle, coherent rotation around a fixed rotation axis within the pulse duration (3 ps), the magnetic field causes Larmor precession (characterized by a frequency ω_L) which forms another, orthogonal, control axis. For any arbitrary superposition between the $|\uparrow\rangle$- and $|\downarrow\rangle$-states, their difference in energy due to the Zeeman energy gives rise to a coherent phase evolution of one spin state compared to the other:

$$|\Psi\rangle = c_\uparrow |\uparrow\rangle + c_\downarrow |\downarrow\rangle$$

$$\begin{pmatrix} c_\downarrow(t+\Delta t) \\ c_\uparrow(t+\Delta t) \end{pmatrix} = \begin{pmatrix} e^{-i\omega_L \Delta t} & 0 \\ 0 & 0 \end{pmatrix} \begin{pmatrix} c_\downarrow(t) \\ c_\uparrow(t) \end{pmatrix} \quad (3.26)$$

Subtracting a common energy scaling factor, we obtain the following evolution operator, which can be written in terms of Pauli spin operators (we change to a reference frame where the magnetic field is supposed to be oriented along z – the extension to our normal frame with the field along x is straightforward):

$$U_{\text{eff}} = \begin{pmatrix} e^{-i\omega_L \Delta t/2} & 0 \\ 0 & e^{+i\omega_L \Delta t/2} \end{pmatrix}$$

$$U_{\text{eff}} = e^{-i\omega_L \Delta t \left(\frac{\sigma_z}{2}\right)} = R_{\vec{B}}(\omega_L \Delta t). \quad (3.27)$$

Hence, the magnetic field determines a second rotation axis, orthogonal to the pulse-induced rotation axis, with the rotation angle determined by the duration of the Larmor precession. The combination of these two coherent control operations can therefore be used to generate any effective rotation by decomposing it into a combination of Larmor precession and pulse-induced rotation [2, 10].

Figure 3.9 illustrates this through so-called Ramsey-interferometry: while a fist pulse creates a particular superposition of the $|\uparrow\rangle$- and $|\downarrow\rangle$-states, the coherent phase evolution due to Larmor-precession will change the phase of the coherent superposition continuously. Depending on the arrival time of another rotation pulse, the net population in the $|\downarrow\rangle$-state (which can be measured with our optical pumping/single photon measurement technique) will oscillate at a rate determined by the Larmor-precession frequency.

For the data in Fig. 3.9a, rotation pulses of $\pi/2$ and π are used. For a $\pi/2$-pulse, the effect of the rotation pulse is so as to rotate the spin into the equator of the Bloch sphere (equal superposition of spin up and spin down). The subsequent Larmor precession then changes the phase of the superposition, after which the final $\pi/2$-pulse brings the spin back into the measurement basis. For a correctly timed second pulse, in phase with the Larmor precession, the spin is rotated into the $|\downarrow\rangle$-state (bright), while for a measurement pulse in antiphase, the spin is rotated to the $|\downarrow\rangle$-state (dark). Hence, in theory, the resulting Ramsey fringes have full visibility.

3.2 All-Optical SU(2) Control

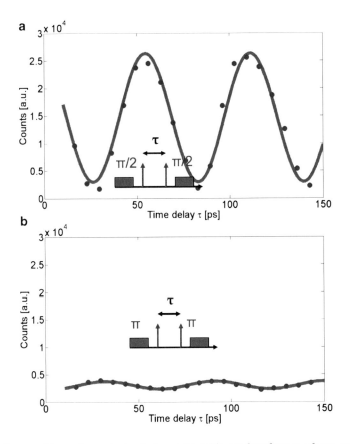

Fig. 3.9 Ramsey fringes for a single electron spin qubit. (**a**) Interference of two $\pi/2$-pulses; (**b**) same, for two π-pulses. *Inset*: pulse sequence used

For Ramsey interference with two π-pulses, the first pulse brings the spin into the $|\downarrow\rangle$-state. Now, Larmor precession only adds a global phase, but does not change the phase of the superposition. The next π-pulse then transforms the spin again into the $|\uparrow\rangle$-state, which is dark. Therefore, π-pulse interference is expected to not yield any visibility of the Ramsey fringes, which is approximately reflected in Fig. 3.9b (small deviations are due to the off-axisness of the pulses: we refer to Ref. [2] and Appendix A.1.2 for further details).

Ramsey-interference can be used to quantify the fidelity of a single coherent rotation pulse: we refer again to Appendix A.1.2 for a detailed analysis. By combining arbitrary pulse energies (arbitrary rotation angle of the pulse-induced rotation) and arbitrary delays between two pulses (arbitrary rotation angle of the Larmor precession), we can realize any coherent spin superposition and therefore any coherent spin rotation: Fig. 3.10 illustrates how both can be combined to cover the entire surface of the Bloch sphere, thereby demonstrating full SU(2)-control.

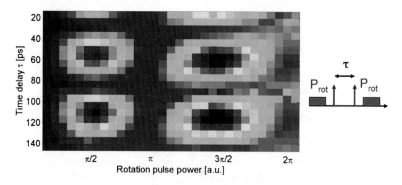

Fig. 3.10 Full control over the surface of the Bloch sphere of a single electron spin qubit, by adjusting the power and the timing of two pulses used in a Ramsey interferometer. *Inset*: pulse sequence used

3.3 Optically Controlled, Quantum Dot Spin Qubits: Coherence

The above description of all-optical, coherent SU(2) control needs to be complemented by the spurious, incoherent interactions of the spin (quantum bit) with the environment. If the spin were a fully isolated object, then all dynamics would be governed by the coherent interaction with the external magnetic field, and the control laser fields. This is, of course, a very crude approximation: the spin is embedded inside a self-assembled quantum dot, a solid-state system that permits interactions with the solid-state environment.

At sufficiently low temperatures, the dominant interaction mechanism involves coupling of the spin qubit to the nuclear spins inside the quantum dot [4,6,7,17–21]. Figure 3.11 schematically illustrates this interaction. As we derived in Sect. 2.1, the electron wavefunction under the effective-mass approximation consists of the product of an envelope wavefunction (blue line) and an underlying, band-wavefunction that reflects the symmetry of the lattice (black lines). For an electron wavefunction, the band-wavefunction has s-type symmetry, which implies that there exists a non-vanishing probability for the electron spin to exist at the core of the constituent lattice sites – exactly the location where the nuclei exist. This is in contrast to the hole-band-wavefunction, which has p-type symmetry, and therefore has a vanishing probability of the hole state to exist at the nuclear core; we refer to Chap. 6 for further details on hole spins.

While other, dipolar interactions also play a (minor) role, the possibility of the electron to exist at the core of the nuclear spins gives rise to a strong contact hyperfine interaction, that adds to the effective Hamiltonian governing the electron spin's dynamics (we use $\vec{S} = \vec{\sigma}/2$):

$$H_{\text{eff}} = H_0 + H_{\text{contact}} \tag{3.28}$$

3.3 Optically Controlled, Quantum Dot Spin Qubits: Coherence

Fig. 3.11 Schematic illustration of the contact hyperfine interaction for electron spins. Due to the s-type symmetry of the underlying wavefunction (*black lines*), the electron spin wavefunction (*blue line*: envelope wavefunction in effective mass approximation – see Sect. 2.1) has a non-vanishing probability amplitude at the location of the nuclear spins (*black arrows*), giving rise to a contact hyperfine interaction characterized by A_{contact}

$$H_0 = g\mu_B \vec{B}\vec{S} \tag{3.29}$$

$$H_{\text{contact}} = \sum_{i=\text{nuclei}} A_{\text{contact},i} \vec{S} \cdot \vec{I}_i. \tag{3.30}$$

Here, $A_{\text{contact},i}$ refers to the contact hyperfine interaction between the electron and the nuclear spin at lattice site i. $\vec{I}_i$ is the nuclear spin at location i: for Ga and As, the nuclear spin is a spin-3/2 particle, while for In, all isotopes are spin-9/2 [19]. By combining terms, we can write the contact hyperfine interaction in terms of the coupling between the electron spin and an effective magnetic field, $\vec{B}_n$:

$$H_{\text{contact}} = \sum_{i=\text{nuclei}} A_{\text{contact},i} \vec{S} \cdot \vec{I}_i \tag{3.31}$$

$$= g\mu_B \left(\frac{\sum_{i=\text{nuclei}} A_{\text{contact},i} \vec{I}_i}{g\mu_B} \right) \cdot \vec{S} = g\mu_B \vec{B}_n \cdot \vec{S} \tag{3.32}$$

$$\vec{B}_n = \frac{\sum_{i=\text{nuclei}} A_{\text{contact},i} \vec{I}_i}{g\mu_B} \tag{3.33}$$

This effective magnetic field adds to the external magnetic field, to yield a net effective magnetic field $\vec{B}_{net} = \vec{B} + \vec{B}_n$. Its effect is two-fold. First, as the nuclear spin's magnetic moment is some three orders of magnitude smaller than the electron's, its energetic splitting under the application of an external field is also about three orders of magnitude smaller. Therefore, at few-K temperatures, one would expect each nuclear spin to be approximately unpolarized (random spin polarization per site). As a consequence, the distribution of the ensemble of nuclear spins inside the quantum dot ($N_{\text{nuclei}} \sim 10^4 - 10^5$ for the self-assembled quantum dots in this work [3]) can be modeled as binomial, without any expected total spin polarization, but with a standard deviation on the order of $\sqrt{N_{\text{nuclei}}}$. This gives rise to a statistical fluctuation of the net effective magnetic field along the direction of the applied field, and therefore also a statistical fluctuation of the Larmor-precession frequency, $\omega_L + \delta\omega_L$.

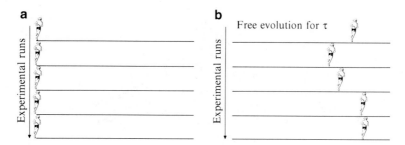

Fig. 3.12 Ensemble dephasing and T_2^*-effects: runners analogy [22]. See text for details

While the nuclear spin fluctuations are slow (typically on ms-timescales [19]), our time-ensemble-based spin-measurement technique is even slower (timescales of seconds to minutes). Therefore, the Larmor-precession frequency can vary from one shot to the other, giving rise to time-ensemble averaging of the Larmor-precession frequency. This effect is schematically illustrated in Fig. 3.12, using the well-known runners analogy [22]. When comparing a single instance of, say, a Ramsey interference experiment to a runner at a particular day of the week, free evolution over a time τ will let her/him advance over a particular distance, determined by his/her particular shape that day. This can be compared with free phase-evolution due to Larmor precession by an amount $\phi_i = \omega_{L,i} \times \tau$. Just as the runner can run different distances on different days for the same amount of time, the phase evolution can differ from shot to shot, depending on the particular Larmor precession frequency, leading to an averaging out of the Ramsey-fringes on a timescale $1/\Delta\omega$. As we shall argue in Sect. 3.3.2, spin-echoes can be used to overcome these ensemble-dephasing effects.

In addition to thermal fluctuations of the nuclear spin ensemble, however, the contact hyperfine interaction can also give rise to induced, electron-spin dependent dynamics of the nuclear spins: this effect is know as dynamic nuclear polarization [4, 6, 7, 19, 21].

3.3.1 Nuclear Spin Interactions

Dynamic nuclear polarization (DNP) effects in self-assembled quantum dots under optical control were observed using spin-locking [20, 23], coherent population trapping [6], resonant absorption in Faraday [7] and Voigt [19] geometry, and in Ramsey interferometry [4]. Figure 3.13, based on data presented in Ref. [4] illustrates the effect of dynamic nuclear spin polarization on Ramsey interferometry using all-optical spin control.

Upon increasing (decreasing) the delay between the $\pi/2$-pulses in a time-averaged Ramsey interferometry experiment, the resulting interference fringes become both asymmetric and hysteretic. As was shown in Ref. [4], this effect can

3.3 Optically Controlled, Quantum Dot Spin Qubits: Coherence

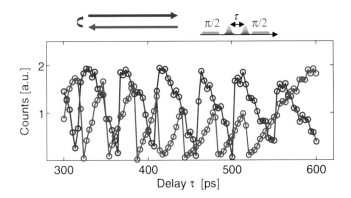

Fig. 3.13 Ramsey-interference fringes for electron spins, showing the effect of dynamic nuclear polarization. When increasing or decreasing the delay between subsequent $\pi/2$-pulses in a Ramsey-interferometry experiment, a non-linear feedback loop with the nuclear spin bath results in asymmetric, hysteretic Ramsey fringes. *Inset*: scan direction (longer, resp. shorter delays) for the different experiments (Experimental data reproduced from [4])

be interpreted as a consequence of the existence of stable manifolds for the nuclear spins, depending on the particular Ramsey-experiment (delay) chosen, and due to the coupled electron/nuclear non-linear dynamics (the non-linearity explains the hysteresis).

Let us consider the interaction between a single electron spin and a single nuclear spin. By rewriting the contact hyperfine coupling in terms of raising and lowering operators [24] (our magnetic field is oriented along the x-direction, which implies raising and lowering operators of the following type: $S^\pm = S_y \pm iS_z, I^\pm = I_y \pm iI_z$), we obtain the following:

$$A_{contact,i}\vec{S}\cdot\vec{I_i} = A_{contact,i}S_xI_x + \frac{A_{contact,i}}{2}(S^+I^- + S^-I^+). \quad (3.34)$$

In the absence of any higher order terms, this contact hyperfine interaction therefore governs the evolution of an individual nucleus.[3] The second term in Eq. 3.34 is the so-called flip-flop term, and indicates that for a nuclear spin to flip, there must be a concurrent and opposite flip of the electron spin – a manifestation of angular momentum conservation.

Figure 3.14 indicates the level structure of an electron-charged, self-assembled quantum dot under full optical control. As the magnetic field splits the electron spin ground states by several 10s of GHz, and the nuclear spins by several 10s of MHz only, this flip-flop mechanism for nuclear spin changes is energetically forbidden to

---

[3] Higher order terms, due to e.g. dipolar coupling to both electron spin and other nuclear spins do exist, and give rise to some slow dynamics on milliseconds- to seconds timescales, as will be shown in Sect. 3.3.2, but for the current analysis, those can be temporarily ignored.

Fig. 3.14 Feedback mechanism giving rise to dynamic nuclear polarization under all-optical electron spin control – see text for details

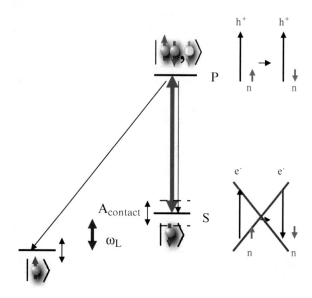

first order: the hyperfine interaction acts as a lock for the nuclear spins. However, whenever the optical pumping cycle excites the system into the $|\uparrow\downarrow\Downarrow\rangle$-state, the nuclear spin dynamics is governed by the interaction with this excited trion state, *not* the contact hyperfine interaction with a single electron spin. This excited state consists of a magnetically inert electron singlet, and an unpaired hole state.

Previously, we described, in Sects. 2.1 and 2.3, the Voigt-geometry trion state $|\uparrow\downarrow\Downarrow\rangle_x$ as superpositions of growth direction trions $(|\uparrow\downarrow\Downarrow\rangle_z \pm |\uparrow\downarrow\Uparrow\rangle_z)$ where the hole states correspond to heavy holes along the growth direction (this was a consequence of a separation of energy scales, where we invoked a large HH-LH splitting to justify working with the HH-mainfold only). While such a description suffices for most of the optical control work described in this dissertation, the hyperfine interaction between the trion state and the nuclear spins requires a slightly refined model. Due to the combined effect of strain and confinement, the hole states along the growth direction are now heavy holes, with a (small) light hole component mixed in (w [6, 9, 19, 25]):

$$|\Uparrow\rangle_z \equiv |J_z = 3/2\rangle + \eta |J_z = -1/2\rangle \tag{3.35}$$

$$|\Downarrow\rangle_z \equiv |J_z = -3/2\rangle + \eta |J_z = 1/2\rangle. \tag{3.36}$$

These HH-LH mixtures (typical values of η for our experiments is around 0.2) now hybridize under the influence of the magnetic field and form the hole-part of the trion states used for optical pumping. As the hole has p-type symmetry, it does not have a contact hyperfine interaction with the nuclear spins; instead, the highest-order hyperfine interaction is dipolar in nature [4, 6]. When written in a basis where the external magnetic field is oriented along the x-direction, and mapping the hole states

3.3 Optically Controlled, Quantum Dot Spin Qubits: Coherence

into a pseudo-spin-1/2 formalism[4] where $|\Uparrow\rangle_z = |S_z = 1/2\rangle, |\Downarrow\rangle_z = |S_z = -1/2\rangle$ such a dipolar interaction takes the following form, characterized by a dipolar-hyperfine constant $A_{\text{dipole},i}$ [6]:

$$H_{\text{dipole},i} = A_{\text{dipole},i}\left[I_{z,i}S_z + O(\eta)(I_{x,i}S_x + I_{y,i}S_y) + O(\eta^2)(I_{z,i}S_x + I_{z,i}S_y)\right] \quad (3.37)$$

The third term in Eq. 3.37 can be written in terms of the raising and lowering operators $I^\pm = I_y \pm iI_z$:

$$(I_{z,i}S_x + I_{z,i}S_y) = \frac{1}{2i}(I_i^+ S_x - I_i^- S_x + I_i^+ S_y - I_i^- S_y). \quad (3.38)$$

These terms allow for flips of the nuclear spin to happen, without any accompanying flip of the hole spin. In other words, when the charged quantum dot is excited into the trion state, the strong lock on the electron spin due to the electron-nuclear contact hyperfine interaction is released – we refer to Fig. 3.14 for a graphical illustration of this effect.

In a Ramsey interferometry or resonant absorption experiment, the probability, C, of creating a trion depends on the state of the electron spin (when the electron spin is in the $|\uparrow\rangle$-state, no trion excitation can occur as the optical pumping laser is resonant with the $|\downarrow\uparrow\Downarrow\rangle$-$|\downarrow\rangle$-transition). Conversely, the state of the electron spin in a Ramsey-interferometry experiment depends on the Larmor precession frequency and therefore on the net nuclear spin polarization. The latter can be characterized in terms of an Overhauser shift y to the Zeeman energy or Larmor precession frequency. The trion creation probability is therefore a function of the delay, τ used in a particular Ramsey interferometry experiment, and the net Overhauser shift, y: $C = C(y, \tau)$.

In addition, for a large net nuclear spin polarization, the energy of the $|\downarrow\rangle$-state will shift to the extent that a fixed-frequency laser will not be able to excite the trion state anymore, adding another Overhauser-dependence to the trion creation rate.

As now the nuclear spin-flip rate depends on the creation of trions, and the creation of trions depends on the nuclear spin polarization, this gives rise to a non-linear feedback loop. In Refs. [4] and [19], we argued that the trion-induced nuclear flip process is the dominant one, followed by a weaker background of nuclear spin diffusion that acts so as to slowly re-equilibrate any net nuclear spin polarization. This feedback process can be modeled in terms of a non-linear drift-diffusion equation for the Overhauser field y [4]:

$$\frac{dy}{dt} = -\kappa y + \alpha \frac{\partial C(y,\tau)}{\partial y} \quad (3.39)$$

---

[4] As the (mixed-in) heavy-holes form an effective two-level system, such re-mapping is valid for the low temperatures in typical experiments: the energy separation from the next higher states is typically several meV.

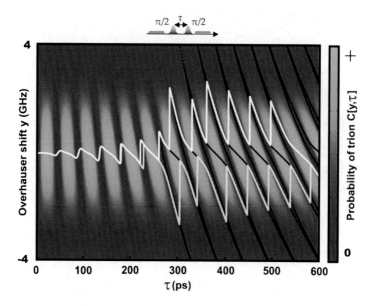

Fig. 3.15 Numerical modelling of the non-linear feedback effects between a single electron spin and nuclear spins, under optical control. *Yellow*: steady-state solutions upon continuously increasing the delay, τ. *White*: solutions for decreasing τ (scanning backwards). See text for details (Figure reproduced from [4])

Here, α and κ are (time-)constants that reflect the effectiveness of the trion-induced nuclear flip process and background nuclear spin diffusion respectively.

Figure 3.15 displays the steady-state solutions of this drift-diffusion equation for the case of Ramsey-interferometry between two $\pi/2$-pulses, with delay τ between them. Due to the non-linearity, the solutions are very much dependent on the initial conditions, which explains the hysteresis that is also observed in Fig. 3.13. For suitable parameters α and κ, the numerical solutions reproduce the experimental results very well – we refer to Appendix B for explicit solutions for both Ramsey-interferometry induced DNP, as well as DNP-effects in resonant absorption experiments (Fig. B.2).

3.3.2 All-Optical Spin-Echo

Even in the absence of DNP-effects, slow changes in the nuclear spin bath affect the results of time-averaged experiments of the electron spin dynamics. As our spin-measurement technique requires many experimental 'shots' in view of the small capture/escape probability of a single photon, each experimental datapoint in e.g. a Ramsey-interferometry experiment requires several seconds of integration – slower than even the higher-order nuclear spin processes which we temporarily ignored

3.3 Optically Controlled, Quantum Dot Spin Qubits: Coherence

before. If the time constant for full thermalization of the nuclear spins is shorter than the integration time of the experiment, then the experiment 'sees' the full width of the nuclear spin distribution. For the self-assembled quantum dots used, the variance in the Larmor-precession frequency can be estimated at about $2\pi \times 100$ MHz [3, 19], resulting in a dephasing, $\Delta\phi = \Delta\omega \times \tau$ on a timescale of a few nanoseconds. This ensemble-averaging time is commonly referred to as T_2^*.

This ensemble-averaging, however, is simply due to our measurement technique, and is not inherent to the spin qubit as such. In order to estimate the usefulness of our spin qubit, the relevant question would be how long a coherent superposition of a single spin would persist, without ensemble averaging. Instead of ensemble averaging over many shots, the phase-evolution of a single instance (in time) of the qubit is now considered. Due to small variations in the environment, fast compared to the duration of a single shot of the experiment, the coherent phase evolution due to e.g. Larmor precession will be affected. Denoting this net, fast variation of the Larmor-precession as $\delta\omega$, we see that the phase now undergoes a random kick, $\delta\phi = \delta\omega \times T$, characterized by a time-constant known as T_2 [3, 23]. It is this single-shot decoherence time that needs to be compared to the duration of a single qubit operation in order to obtain a figure of merit for the qubit.

In order to observe the single-shot decoherence in a multi-shot experiment,[5] one can use the well-known spin-echo technique that was developed by Erwin Hahn for use in NMR-experiments (see Ref. [27]. Ref. [28] describes an early variant that is more robust against pulse errors). The basic idea is illustrated in Fig. 3.16, again in terms of the runners analogy [22].

When applying an optical π pulse along the z-direction, perpendicular to the orientation of the magnetic field, the effect on the spin states (modulo overall phase factors) is the following:

$$R_z(\pi)|\uparrow\rangle_x = |\downarrow\rangle_x \tag{3.40}$$

$$R_z(\pi)|\downarrow\rangle_x = |\uparrow\rangle_x. \tag{3.41}$$

In other words, a π pulses reverses the role of the up- and down spin states. If now, due to ensemble dephasing, the $|\downarrow\rangle_x$-state develops a phase $\Delta\omega \times \tau$ compared to the $|\uparrow\rangle_x$-state, the reversal of up- and down after a π-pulse will cause the other state to develop an equal and opposite phase factor after another time τ. This is equivalent, in the runners analogy, to each day having the runner reverse course, exactly halfway (τ), and return home. Of course, after running full-time (2τ), the runner will always arrive in the 'home' position (Fig. 3.16d), regardless of which day of the week. This effect is known as refocusing due to spin-echo, and the single π-pulse is therefore often referred to as a refocusing pulse [22].

---

[5]Rather: *approximately* observe the single-shot dynamics in a multishot experiment. The spin-echo results in slightly different dynamics for the nuclear spin bath than true, single-shot free induction decay, and therefore, slightly different decoherence effects. We refer to Refs. [5, 18] and [26] for further details.

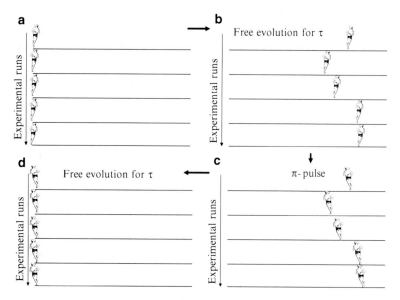

Fig. 3.16 Ensemble dephasing, and spin-echo as a way of overcoming it and measuring T_2: runners analogy [22]. See text for details

As a spin echo refocuses the ensemble dephasing, it is now only sensitive to small fluctuations within a single shot (duration: 2T) of the experiment: $\delta\phi = \delta\omega \times 2T$. Such single-shot decoherence can be studied by a modified form of Ramsey-interferometry, in which a refocusing π-pulse is inserted halfway in between the two $\pi/2$-pulses. Figure 3.17 illustrates the resulting Ramsey-fringes, for slight imbalances, 2τ, between the time before and after the refocusing pulse. By measuring the visibility of the echo-fringes as a function of the total time, $2T$ of the experimental run, the characteristic decay of the coherence can be measured. For all-optical control of a single spin qubit, this was first reported in Ref. [3], where we obtained a T_2-time of 2–3 µs (Fig. 3.17, right panel), characterizing an exponential decay process.

Such a T_2-time needs to be compared to the duration of a single quantum bit operation. The ultrafast pulses used require about 3 ps, and the Larmor-precession time varies between 20 and 60 ps. As any state on the Bloch sphere can be reached within at most half a Larmor precession period after applying an ultrafast pulse, the maximum gate operation time is on the order of 30 ps. Compared to a T_2-time of 2–3 µs, a figure of merit of about 10^5 would be obtained, suggesting that 10^5 gate operations could be applied before decoherence due to variations in the Larmor-precession frequency. Obviously, this is not the whole story: the coherent rotation pulses are imperfect, and have finite fidelities (on the order of 95–99 %, see Appendix A.1.2). Therefore, the application of 10^5 ultrafast coherent control pulses would result in a complete loss of all coherence. However, if the fidelity of a single ultrafast pulse operation could be improved further, then 10^5 operations could be applied before decoherence occurs.

3.3 Optically Controlled, Quantum Dot Spin Qubits: Coherence

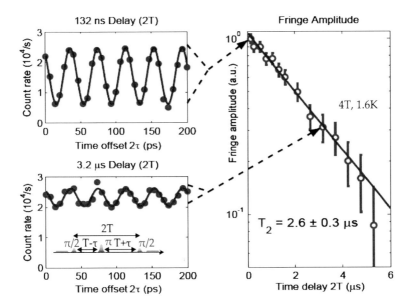

Fig. 3.17 All-optical spin echo for a single electron spin: a single π-pulse refocuses the shot-to-shot variances in the Larmor precession frequency. By measuring the visibility of the resulting Ramsey fringes for slightly imbalanced (net difference: 2τ) spin precessions before resp. after application of the refocusing pulse (*top, bottom left*), the true, spin-echo decoherence can be extracted (*right*). For a single electron spin, this decoherence occurs on timescales around 2–3 μs (Figure reproduced from [3])

The spin-echo technique can also be used to directly visualize T_2^*: as was shown in Refs. [3] and [19], dynamic nuclear polarization effects are much less pronounced when applying spin-echoes (without spin-echo, the DNP effects completely dominate the spin dynamics: we refer to Fig. 3.13). In Fig. 3.18a, spin-echo Ramsey fringes are shown for relatively large differences, $\delta\tau$ between the two halves of the spin-echo experiment.

The data in Fig. 3.18a were obtained using a few seconds of integration per datapoint, after which an electronically controllable stage was slowly moved to the next position. This integration time turned out to be slightly shorter than the thermal equilibration time of the nuclei (some 10 s), such that for each experimental point, only a subset of the full nuclear spin distribution was sampled. This effect is schematically illustrated in Fig. 3.18b. As we sample a narrower subdistribution per datapoint, the decay of the envelope function of the Ramsey-fringes seemingly occurs on a slower timescale:

$$\int_{subset,j} e^{-i(\omega+\Delta\omega)t} F_{subset,j}(\Delta\omega) d\Delta\omega$$
$$\sim e^{-i(\omega+\Delta\omega_{j,0})t} e^{-(\Delta\omega_{subset,j}t)^2} \tag{3.42}$$

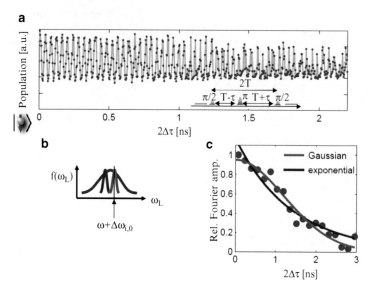

Fig. 3.18 T_2^*-decay, visualized by means of a spin echo. See text for details (Figure taken from [3])

(we assume a Gaussian distribution for each subset, characterized by $\Delta\omega_{subset}$). However, the *phase* of each datapoint, $(\omega + \Delta\omega_{j,0})t$ is now taken from the full nuclear spin distribution (the nuclear spin equilibration time is faster than the duration of the entire experiment, involving stage movements and integration for several datapoints). We can therefore notice a phase-randomization in the data in Fig. 3.18 before any amplitude-effects. Quantifying that phase-randomization by a running Fourier-transformation, we see that the decay due to the full nuclear spin distribution is now characterized by another Gaussian, with a time constant on the order to 1.5 ns (Fig. 3.18c).

References

1. J. Berezovsky, M. H. Mikkelsen, N. G. Stoltz, L. A. Coldren, and D. D. Awschalom. Picosecond coherent optical manipulation of a single electron spin in a quantum dot. *Science*, 320:349, 2008.
2. D. Press, T. D. Ladd, B. Zhang, and Y. Yamamoto. Complete quantum control of a single quantum dot spin using ultrafast optical pulses. *Nature*, 456:218, 2008.
3. D. Press, K. De Greve, P. McMahon, T. D. Ladd, B. Friess, C. Schneider, M. Kamp, S. Höfling, A. Forchel, and Y. Yamamoto. Ultrafast optical spin echo in a single quantum dot. *Nat. Photonics*, 4:367, 2010.
4. T. D. Ladd, D. Press, K. De Greve, P. McMahon, B. Friess, C. Schneider, M. Kamp, S. Höfling, A. Forchel, and Y. Yamamoto. Pulsed nuclear pumping and spin diffusion in a single charged quantum dot. *Phys. Rev. Lett.*, 105:107401, 2010.

References

5. W. M. Witzel and S. Das Sarma. Quantum theory for electron spin decoherence induced by nuclear spin dynamics in semiconductor quantum computer architectures: Spectral diffusion of localized electron spins in the nuclear solid-state environment. *Phys. Rev. B*, 74:035322, 2006.
6. X. Xu *et al*. Optically controlled locking of the nuclear field via coherent dark-state spectroscopy. *Nature*, 459(4):1105, 2009.
7. C. Latta *et al*. Confluence of resonant laser excitation and bidirectional quantum-dot nuclear-spin polarization. *Nat. Phys.*, 5:758, 2009.
8. S. M. Clark, K-M. C. Fu, T. D. Ladd, and Y. Yamamoto. Quantum computers based on electron spins controlled by ultrafast off-resonant single optical pulses. *Phys. Rev. Lett.*, 99:040501, 2007.
9. K. De Greve, P. L. McMahon, D. Press, T. D. Ladd, D. Bisping, C. Schneider, M. Kamp, L. Worschech, S. Höfling, A. Forchel, and Y. Yamamoto. Ultrafast coherent control and suppressed nuclear feedback of a single quantum dot hole qubit. *Nat. Phys.*, 7:872, 2011.
10. M. A. Nielsen and I. L. Chuang. *Quantum Computation and Quantum Information*. Cambridge University Press, 2000.
11. J Berezovsky, M. H. Mikkelsen, O. Gywat, N. G. Stoltz, L. A. Coldren, and D. D. Awschalom. Nondestructive Optical Measurements of a Single Electron Spin in a Quantum Dot. *Science*, 314:1916, 2006.
12. M. Atatüre, J. Dreiser, A. Badolato, and A. Imamoglu. Observation of Faraday rotation from a single confined spin. *Nat. Phys.*, 3:101, 2007.
13. A. N. Vamıvakas, C.-Y. Lu, C. Matthiesen, Y. Zhao, S. Fält, A. Badolato, and M. Atatüre. Observation of spin-dependent quantum jumps via quantum dot resonance fluorescence. *Nature*, 467:297, 2010.
14. X. Xu, Y. Wu, B. Sun, Q. Huang, Jun Cheng, D. G. Steel, A. S. Bracker, D. Gammon, C. Emary, and L. J. Sham. Fast spin state initialization in a singly charged InAs-GaAs quantum dot by optical cooling. *Phys. Rev. Lett.*, 99:097401, 2007.
15. K.-M. C. Fu *et al*. Ultrafast control of donor-bound electron spins with single detuned optical pulses. *Nat. Phys.*, 4:780, 2008.
16. V. N. Golovach, A. Khaetskii, and D. Loss. Phonon-Induced Decay of the Electron Spin in Quantum Dots. *Phys. Rev. Lett.*, 93:016601, 2004.
17. D. V. Bulaev and D. Loss. Spin decoherence and relaxation of holes in a quantum dot. *Phys. Rev. Lett.*, 95:076805, 2005.
18. W. A. Coish and D. Loss. Hyperfine interaction in a quantum dot: Non-Markovian electron spin dynamics. *Phys. Rev. B*, 70:195340, 2004.
19. T. D. Ladd, D. Press, K. De Greve, P. McMahon, B. Friess, C. Schneider, M. Kamp, S. Höfling, A. Forchel, and Y. Yamamoto. Nuclear feedback in a single quantum dot under pulsed optical control. arXiv:1008.0912v1.
20. A. Greilich *et al*. Nuclei-induced frequency focusing of electron spin coherence. *Science*, 317(4):1896, 2007.
21. I. T. Vink *et al*. Locking electron spins into magnetic resonance by electron–nuclear feedback. *Nat. Phys.*, 5:764–768, 2009.
22. L. Allen and J. H. Eberly. *Optical Resonance and Two-level Atoms*. Dover books on Physics, 1987.
23. Greilich, A *et al*. Mode locking of electron spin coherences in singly charged quantum dots. *Science*, 313:341, 2006.
24. A. Messiah. *Quantum mechanics*. Dover, 1999.
25. J. Fischer and D. Loss. Hybridization and Spin Decoherence in Heavy-Hole Quantum Dots. *Phys. Rev. Lett.*, 105:266603, 2010.
26. R.-B. Liu, W. Yao, and L. J. Sham. Control of electron spin decoherence caused by electron-nuclear spin dynamics in a quantum dot. *New J. Phys.*, 9:226, 2007.
27. E. L. Hahn. Spin Echoes. *Phys. Rev.*, 80:580, 1950.
28. H. Y. Carr and E. M. Purcell. Effects of Diffusion on Free Precession in Nuclear Magnetic Resonance Experiments. *Phys. Rev.*, 94:630, 1954.

Chapter 4
All-Optical Hadamard Gate: Direct Implementation of a Quantum Information Primitive

The ultrafast optical spin control techniques described in Chap. 3 were derived under the assumption of the pulses being short compared to the Larmor precession frequency. In practice, however, picosecond-timescale pulses are used, combined with Larmor precession periods of several tens of picoseconds [1,2]. While lowering the magnetic field and increasing the Larmor precession period would render the picosecond pulses (relatively) 'shorter', this would come at the expense of the speed of elementary quantum gates (the Larmor precession acts as one of the fundamental 1-qubit gates). In addition, decreasing the magnetic field lowers the spin readout fidelity, as the filtering of the optical pumping laser becomes more troublesome – we refer to Sect. 3.2.1 and Appendix A.1.1.

In principle, reducing the pulse duration by moving to the femtosecond regime could be an option, provided that the detuning is adjusted accordingly to compensate for the much broader bandwidth of the femtosecond pulses. However, in the studies reported in Refs. [1,2], it was found that shorter pulse durations, in view of the larger peak powers, result in reduced pulse fidelities. The exact origin of this phenomenon is currently unknown, but is likely due to spurious interactions with the solid-state environment (the description of the charged, self-assembled quantum dot as a simple four-level structure is a crude one, after all).

Hence, the timescales of the coherent optical control pulses and the magnetic field-induced Larmor precession are within the same order of magnitude, which requires a more advanced description. In particular, the competition between the finite-duration pulse-induced coherent rotation and the Larmor precession will be shown to result in effectively off-axis pulses [1], which particularly affects the optical π-pulses used in refocusing schemes such as the optical spin-echo (see Sect. 3.3.2).

4.1 Finite Pulse Duration: Off-Axis Spin Rotations

Figure 4.1 revisits the Rabi-oscillations which we previously discussed in Fig. 3.8. The count rate, and therefore the probability of finding the system in the $|\downarrow\rangle$-state, is markedly different for a π-pulse than for a 3π-pulse. As first reported in Ref. [1], the origin thereof is in a competition between the magnetic-field induced Larmor precession (effective rotation around an axis parallel to the magnetic-field-direction in the lab-frame) and the optically-induced spin rotation (around an axis parallel to the Poynting vector of the incoming light field).

In the following, we will use the axis convention previously discussed in Fig. 3.5b: the magnetic field is oriented along the x-axis, and the optical control occurs along the z-axis (in the Voigt-geometry, this is the growth axis). By transforming Eq. 3.5 accordingly, we obtain:

$$\frac{H_{\text{eff}}}{\hbar} = \begin{pmatrix} \frac{\Omega_{\text{eff}}(t)}{2} & \frac{\delta_e}{2} \\ \frac{\delta_e}{2} & -\frac{\Omega_{\text{eff}}(t)}{2} \end{pmatrix} \quad (4.1)$$

$$\Omega_{\text{eff}}(t) \sim \frac{|\Omega(t)|^2}{4\Delta} \quad (4.2)$$

In Eq. 4.2, Δ refers to the detuning from the excited trion states, while $\Omega(t)$ refers to the spin-trion coupling due to the fast, time-dependent optical pulse. Assuming, temporarily, an optical coupling field that is *constant* in time,[1] we can rewrite this in terms of Pauli spin-operators:

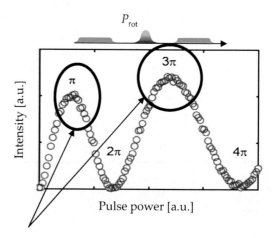

Fig. 4.1 Off-axis rotation pulses resulting in imperfect Rabi-oscillations; see text for details

---

[1]This would correspond to a coupling field that can be switched on and off infinitely fast, which is of course non-physical ...

4.1 Finite Pulse Duration: Off-Axis Spin Rotations

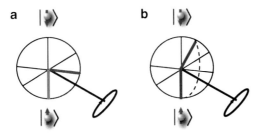

Fig. 4.2 Finite-duration, off-axis rotation pulses. (**a**) Combined effect of the Larmor precession and the optical spin manipulation, resulting in an effective, off-axis rotation. (**b**) Effect of an off-axis π pulse on a spin-eigenstate. The off-axis pulse does not reverse the role of $|\uparrow\rangle$ and $|\downarrow\rangle$, resulting in imperfect refocusing in e.g. spin-echo sequences. See text for details

$$\frac{H_{\text{eff}}}{\hbar} = \Omega_{\text{eff}} \frac{\sigma_z}{2} + \delta_e \frac{\sigma_x}{2} \qquad (4.3)$$

From Eq. 4.3, we see that for a pulse duration t_{eff} we would obtain the following evolution operator, U_{eff}:

$$U_{\text{eff}} = e^{-it_{\text{eff}}\sqrt{\Omega_{\text{eff}}^2 + \delta_e^2}\left(\sigma_z \times \frac{\Omega_{\text{eff}}}{2\sqrt{\Omega_{\text{eff}}^2 + \delta_e^2}} + \sigma_x \times \frac{\delta_e}{2\sqrt{\Omega_{\text{eff}}^2 + \delta_e^2}}\right)} \qquad (4.4)$$

$$U_{\text{eff}} = e^{-i\theta_{\text{net}}\left(\frac{\vec{\sigma}\cdot\vec{n}_{\text{net}}}{2}\right)} = R_{\vec{n}_{\text{net}}}(\theta_{\text{net}})$$

$$\theta_{\text{net}} = \sqrt{\Omega_{\text{eff}}^2 + \delta_e^2} \times t_{\text{eff}} = \Omega_{\text{net}} \times t_{\text{eff}}$$

$$\vec{n}_{\text{net}} = \left(\frac{\delta_e}{\Omega_{\text{net}}}, 0, \frac{\Omega_{\text{eff}}}{\Omega_{\text{net}}}\right) \qquad (4.5)$$

This situation is schematically illustrated in Fig. 4.2a: the combination of the rotation pulse and the Larmor precession results in a rotation around an effective axis, $\vec{n}_{\text{net}}$, that is the weighted vector-sum of the Larmor- and optical rotation axis. For the more realistic case of a time-varying laser field with e.g. a Gaussian or hyperbolic-secant envelope, the analysis is slightly more complex: as σ_x and σ_z do not commute, evaluation of the evolution operator requires some care – we refer to Ref. [1] for a full, numerical simulation. The net effect, however, is the same for either constant or time-varying laser fields: an off-axis rotation, along an axis that is a weighted vector sum of the Larmor- and optical rotation axis. For an optical π-pulse around this effective axis, as indicated in Fig. 4.2b, the net effect is that, for initialization at the bottom of the Bloch sphere, a π-pulse does not transfer the state to the top of the Bloch sphere, and vice versa. Our naive analysis also explains the difference in height between a π- and a 3π-pulse in Fig. 4.1: as we obtain the Rabi-oscillations by changing the power of the rotation pulse, the effective rotation axis changes, approaching more and more the optical rotation axis for large optical powers. Unfortunately, large powers also result in large incoherent effects, as was shown in Ref. [1].

As we shall demonstrate in Sect. 4.3, the effect of the off-axis rotation is particularly detrimental for optical π-pulses, which are used for refocusing in spin-echo sequences. As the net, off-axis π pulses now do not anymore switch the role of the $|\uparrow\rangle$- and $|\downarrow\rangle$-states, this refocusing is now imperfect, which results in a loss of coherence (visibility) in spin-echo experiments.

4.2 Composite Pulses: Hadamard Gates

π-rotation pulses around the optical rotation axis ($R_z(\pi)$) cannot be realized by an optical rotation pulse alone; however, it is possible to compose an effective π-pulse by compounding several pulses. One such composition is the well-known $\sigma_z = H\sigma_x H$-decomposition: here, $\sigma_x = iR_x(\pi)$ refers to a π-rotation around the magnetic field axis, which can be realized by Larmor precession, while H stands for the Hadamard gate [3], and $\sigma_z = iR_z(\pi)$. Therefore, $R_z(\pi) = HR_x(\pi)H$, as indicated in Fig. 4.3b. The Hadamard gate is perhaps the most important of the single-qubit gates, and was already briefly discussed in Sect. 1.1.3.2. Its effect is to map the spin eigenstates (we work in the basis of the magnetic field), $|\downarrow\rangle$ and $|\uparrow\rangle$, into two orthogonal, equal-weight superpositions of the spin states – in other words, into two orthogonal states on the equator of the Bloch-sphere:

$$H(|\downarrow\rangle) = \frac{1}{\sqrt{2}}(|\uparrow\rangle - |\downarrow\rangle) \tag{4.6}$$

$$H(|\uparrow\rangle) = \frac{1}{\sqrt{2}}(|\uparrow\rangle + |\downarrow\rangle) \tag{4.7}$$

The Hadamard gate can be written in terms of Pauli-spin operators as follows:

$$H = \frac{\sigma_x + \sigma_z}{\sqrt{2}} \tag{4.8}$$

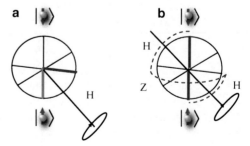

Fig. 4.3 (a) Hadamard gate, a fundamental primitive of quantum information. (b) A composite π-pulse, based on two Hadamard gates and a Larmor-precession (Z-gate)

4.2 Composite Pulses: Hadamard Gates

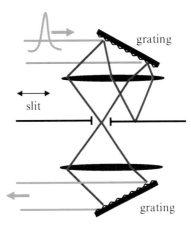

Fig. 4.4 A grating-based pulse stretcher. See text for details

We can now invoke the following identity (see Ref. [3]),

$$R_{\vec{n}}(\theta) = \cos(\frac{\theta}{2})I - i\sin(\frac{\theta}{2})(n_x\sigma_x + n_y\sigma_y + n_z\sigma_z) \quad (4.9)$$

from which we find the following realization of a Hadamard gate:

$$H = ie^{-i\pi(\frac{\sigma_x + \sigma_z}{2\sqrt{2}})} \quad (4.10)$$

Equation 4.10 is, to within a global phase factor, a rotation by an angle π around an effective axis that is an equal-weight combination of Larmor-precession and pulse-induced rotation – similar to the case described in Eq. 4.5. In other words: for the simplistic case of constant pulse amplitude, matching the effective pulse-Rabi-frequency Ω_{eff} to the Larmor-precession δ_e and adjusting the pulse duration such that $t_{\text{eff}} \times \sqrt{2}|\Omega_{\text{eff}}| = \pi$ will result in an effective rotation that is just enough off-axis to generate a Hadamard gate. For truly time-varying pulses, the analysis is more complex, but the final conclusion is the same: by adjusting the pulse amplitude *and* duration in a two-dimensional (numerical, or heuristic) search, an effective Hadamard gate can be created.

While modulating the amplitude of the rotation pulse is straightforward, adjusting its duration is less trivial. Our picosecond rotation pulses are generated by an actively modelocked laser (Spectra-Physics Tsunami), with limited tunability of the pulse duration. Instead, we use a double-grating based pulseshaper [4–6] as displayed in Fig. 4.4. Such a grating shaper is classic example of Fourier-optics: for a Fourier-transform-limited pulse, the first grating disperses the pulse based on its wavevector ($\vec{k}$) composition. The first lens, one focal length removed from both the grating and the image plane, acts as a Fourier transformation from $\vec{k}$-space to real space: different wavevectors are imaged in different locations. Therefore, spatial modulation within the image-plane corresponds to wavevector-modulation (and, for transform-limited, unchirped pulses also frequency-modulation). A second lens and grating respectively transform back to $\vec{k}$-space and recombine the respective wavevectors into an unchirped pulse which is now temporally modulated by virtue of the spatial (frequency) modulation.

The simplest possible modulation consists of cutting the pulse: such a restriction in frequency will result in an effective stretching in time, as both are related by a Fourier-transformation. While more sophisticated modulation schemes exist, based e.g. on spatial light modulators, we find that a simple slit provides enough modulation to realize reasonable pulse stretching.

4.3 Composite Π-Pulses: Spin-Echo and Refocusing

Using a slit and a grating shaper, we perform a heuristic, 2-dimensional search in order to obtain the best Hadamard gate. We use the visibility of the spin-echo-Ramsey-fringes as metric. The results are summarized in Figs. 4.5 and 4.6. We use a magnetic field of 9 T, corresponding to a Larmor-precession frequency of 20 ps. For a pulse duration of 3–4 ps, the off-axis π-pulses affect the visibility of the spin-echo fringes significantly: in Fig. 4.6a, the visibility is around 60%. By cutting the pulse to a duration of some 5–6 ps, verified with an autocorrelation setup (FROG, SwampOptics) and optimizing the pulse power accordingly, we obtain a Hadamard pulse that significantly improves the echo visibility and fidelity.

Figure 4.5 shows Ramsey-fringes using such Hadamard pulses. We realize our effective π-pulse by adjusting the delay between the Hadamard pulses: we look for a maximum in the Ramsey-interferometer. We then fix the delay between these Hadamard pulses, and use this composite pulse as an effective π-pulse in a spin-echo experiment (in comparison with the setup outlined in Fig. 3.7, this requires splitting off a third path from the modelocked laser, with variable delay and individual pulse control by a free-space electro-optic modulator).

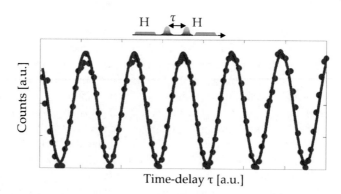

Fig. 4.5 Ramsey interferometry, using two Hadamard gates and a variable delay, τ (Larmor-precession). See text for details. *Inset*: pulse sequence used

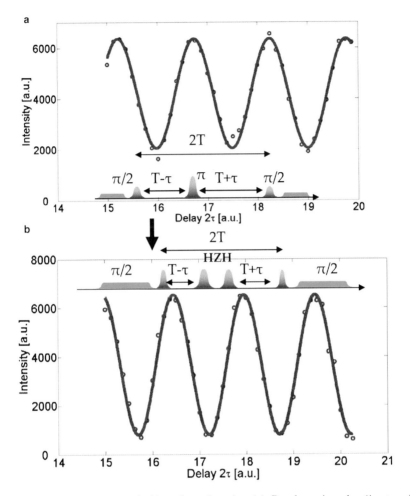

Fig. 4.6 Hadamard gates as primitives for spin echo. (**a**) Regular spin echo (*inset*: pulse sequence used), with limited visibility due to an off-axis π-pulse. (**b**) Spin-echo, using composite, Hadamard-gate based spin echo (*inset*: pulse sequence used). The visibility is significantly increased

In Fig. 4.6b, we see that the resulting echo-fringes now have significantly improved visibility: upwards of 90%, confirming the effectiveness of our Hadamard gates.

References

1. D. Press, T. D. Ladd, B. Zhang, and Y. Yamamoto. Complete quantum control of a single quantum dot spin using ultrafast optical pulses. *Nature*, 456:218, 2008.

2. D. Press, K. De Greve, P. McMahon, T. D. Ladd, B. Friess, C. Schneider, M. Kamp, S. Höfling, A. Forchel, and Y. Yamamoto. Ultrafast optical spin echo in a single quantum dot. *Nat. Photonics*, 4:367, 2010.
3. M. A. Nielsen and I. L. Chuang. *Quantum Computation and Quantum Information*. Cambridge University Press, 2000.
4. A. M. Weiner. Femtosecond pulse shaping using spatial light modulators. *Rev. Sci. Instrum.*, 71:1929, 2000.
5. Eugene Hecht. *Optics (2nd ed.)*. Addison Wesely, 1987.
6. Joseph W. Goodman. *Introduction to Fourier Optics*. Roberts and Company Publishers, 2005.

Chapter 5
Fast, Pulsed, All-Optical Geometric Phases Gates

In the previous chapters, detuned optical pulses were used to generate qubit rotations. While deriving the effect of the laser pulses on the spin, we generally ignored overall, global phases. However, in this chapter, we shall report on experiments were the global (geometric) phases of a (detuned) 2-level interaction can be visualized in Ramsey interferometry, by beating the phase against the Larmor precession. In the context of 2-qubit interactions, geometric phases play a crucial role for the realization of entangling gates [1, 2]; the realization of a fast, pulsed global phase for single quantum dot electron spins is essential for this task.

5.1 Global Phase of a 2-Level System upon a Cyclic Transition

Let us consider a generic two-level system, coupled by a detuned (detuning: Δ) optical interaction characterized by a Rabi-frequency Ω_P, as in Fig. 5.1. Denoting the excited state by $|\uparrow\downarrow\Downarrow\rangle$ (with energy E_{exc}) and the ground state by $|\downarrow\rangle$, we obtain the following set of equations after application of a rotating-wave approximation:

$$|\Psi_{2-\text{level}}\rangle = a(t)e^{-iE_{exc}t/\hbar}|\uparrow\downarrow\Downarrow\rangle + b(t)|\downarrow\rangle$$

$$i\begin{pmatrix}\dot{a}(t)\\\dot{b}(t)\end{pmatrix} = \begin{pmatrix}\Delta & -\frac{\Omega_P}{2}\\-\frac{\Omega_P^*}{2} & 0\end{pmatrix}\begin{pmatrix}a(t)\\b(t)\end{pmatrix}. \quad (5.1)$$

This interaction Hamiltonian can be written as follows, in terms of Pauli-spin operators, σ_i and the unit-tensor I (we assume Ω_P to be real; the case for a complex Ω_P simply changes the orientation of the effective rotation axis in the following, but does not fundamentally change the solutions):

$$H_{int}/\hbar = \frac{\Delta}{2}I + \frac{\Delta}{2}\sigma_z - \frac{\Omega_P}{2}\sigma_x \quad (5.2)$$

Fig. 5.1 Schematic diagram of the detuned interaction with a 2-level system, used for derivation of the global phase upon cycling through the transition. Δ detuning, Ω_p Rabi frequency of the excitation

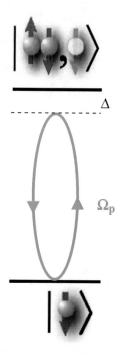

Using Eq. 5.2, the evolution operator, $U = e^{-iH_{\text{int}}t/\hbar}$, can now be written as:

$$U = e^{-i\frac{\Delta}{2}t} \times e^{+i\Omega_{\text{net}}t\left(\frac{\vec{\sigma}\cdot\vec{n}_{\text{net}}}{2}\right)} \tag{5.3}$$

$$= e^{-i\frac{\Delta}{2}t} \times R_{\vec{n}_{\text{net}}}(\Omega_{\text{net}}t) \tag{5.4}$$

$$\Omega_{\text{net}} = \sqrt{\Delta^2 + |\Omega_P|^2} \tag{5.5}$$

$$\vec{n}_{\text{net}} = \left(\frac{\Omega_P}{\Omega_{\text{net}}}, 0, -\frac{\Delta}{\Omega_{\text{net}}}\right) \tag{5.6}$$

Equation 5.4 has the form of an overall phase factor, compounded with a rotation around an axis, $\vec{n}_{\text{net}}$. When applying a 2π rotation, the net effect of this rotation is to add a π phase factor. We therefore have that, for a 2π cycling transition:

$$t_{\text{cycle}} = \frac{2\pi}{\Omega_{\text{net}}} \tag{5.7}$$

$$\phi_{\text{net}} = \pi - \frac{\Delta}{2} \times t_{\text{cycle}}$$

$$= \pi \times \left(1 - \frac{\Delta}{\sqrt{\Delta^2 + |\Omega_P|^2}}\right) \tag{5.8}$$

The expression for the net, global phase in Eq. 5.8 is valid for any detuning and Rabi-frequency, provided that a full cycling transition is realized (i.e., $\Omega_{net} \times t_{net} = 2\pi$): this condition is only fullfilled for particular durations t_{net} of the coupling field.

For a far-detuned transition, however, and for sufficiently slowly evolving and weak Rabi-frequencies, the adiabatic theorem [3] can be invoked, where for *any* pulse duration the system will remain in the ground state, and the net phase can be described in terms of the AC-Stark effect as derived in Sect. 3.1.2:

$$\phi_{net,adiabatic} = \frac{\Delta}{2}\left(\sqrt{1 + \frac{|\Omega_P|^2}{\Delta^2}} - 1\right) \times t_{net,adiabatic}$$

$$\simeq \frac{|\Omega_P|^2}{4\Delta} \times t_{net,adiabatic} \qquad (5.9)$$

Equation 5.9 assumes a coupling field that is constantly on – for the more general case of a laser field with a particular temporal shape $\Omega_P(t)$, we refer to Refs. [4, 5]. In particular, an analytic solution for a 2π hyperbolic-secant pulse shape (sech(σt)) results in [5]:

$$\phi_{net,hyp-sec} = 2\arctan\left(\frac{\sigma}{\Delta}\right) \qquad (5.10)$$

Note that this expression is an implicit function of the pulse strength (Ω_P), enforced through the condition for a 2π-pulse.

For an on-resonance laser field, going through a cyclic transition will therefore change the global phase by a phase-factor of π.

5.2 Visualizing the Global Phase: Ramsey Interferometry

In itself, this global phase is generally irrelevant, as it can be taken up by a change of the time-origin in the free-evolution. It is only when compared to another state which is *not* subject to the same global phase that things become interesting. One such example occurs in the case of Ramsey-interferometry [6, 7], as indicated in Fig. 5.2. As we reported in Sect. 3, the first $\pi/2$ pulse in a Ramsey interferometer will realize a superposition of spin eigenstates that is located in the equator of the Bloch sphere, which will evolve due to Larmor-precession.

If, however, one of the states in the superpositions undergoes a cyclic transition (the $|\downarrow\rangle$-state in Fig. 5.2 is coupled to the $|\uparrow\downarrow\Downarrow\rangle$-trion state in Fig. 5.2), then after one such cycle that state will undergo a phase shift of π, on top of the phase factor developing due to Larmor precession. In the Ramsey-interferometer, therefore, one would expect to see a π phase shift after such a cyclic transition. This effect was observed, for CW coupling to a trion state, in Ref. [7].

A CW field, however, is in practice not very useful: only for particular times $t_{cycle} = 2\pi/\Omega_P$ will the $|\downarrow\rangle$-$|\uparrow\downarrow\Downarrow\rangle$ 2-level-system return to the ground state. For any other time, it will end up in some superposition of the spin- and the trion-state, which

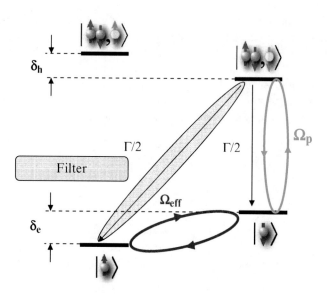

Fig. 5.2 Schematic of the 4-level structure used for Ramsey interferometry, in which the global phase of one spin state due to a cyclic optical transition can be visualized with regard to the other spin state. See text for details

is outside the Hilbert space of the qubit. Moreover, as the trionstate can decay by spontaneous emission, the sheer possibility of this decay would provide which-path information that would reduce the visibility of the interference fringes.

5.3 Pulsed, Fast Geometric Phase of a Single Electron Spin Qubit

We therefore implement a fast, pulsed version of the geometric/global phase, in order to realize an actual gate. The laser field coupling the $|\downarrow\rangle$- and $|\uparrow\downarrow\Downarrow\rangle$-states is derived by modulating a narrowband, CW-diodelaser. The modulation is provided by a fiber-based electro-optic modulator (EOSpace), which is in principle capable of operating up to 20 GHz or higher. The modulation is driven by a 10 GB/s pulse-pattern generator (PPG, Anritsu), which is synchronized to the ultrafast rotation laser used for creating the coherent spin superpositions by means of a high-frequency phase-locked loop (PLL, 32nd harmonic of the 75.76 MHz repetition rate, Valon Technologies) and an electronic quadrupler (Marki Microwave). The Ramsey-interference experiment itself was extensively described in Chap. 3 – other than the fast modulation of the CW-laser, the techniques and setups used are the same as those reported there [6]. In the experiment, an approximately

5.3 Pulsed, Fast Geometric Phase of a Single Electron Spin Qubit

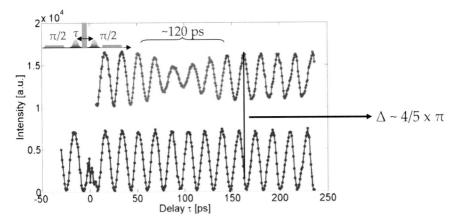

Fig. 5.3 Experimental Ramsey interferometry data, upon a cyclic $|\downarrow\rangle - |\uparrow\downarrow\Downarrow\rangle$-transition of ~120 ps, due to modulation of a resonant CW laser by an electro-optic modulator. The net phase shift is around $4/5\pi$. *Inset*: optical pulse scheme used. See text for details

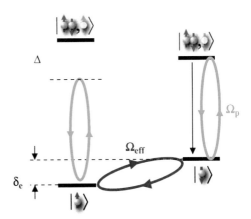

Fig. 5.4 In an electron-charged quantum dot in the Voigt geometry, the on-resonant phase of one spin branch is partially compensated by the detuned phase of the other spin branch, explaining the data in Fig. 5.3

100 ps pulse, on resonance with the $|\downarrow\rangle$-$|\uparrow\downarrow\Downarrow\rangle$-transition and with appropriate power is applied in between two $\pi/2$-pulses; this case is then compared to the Ramsey interference without such geometric-phase generating pulse.

The data are shown in Fig. 5.3. The blue trace shows the Ramsey interference fringes without the intermediate excitation of the $|\downarrow\rangle$-$|\uparrow\downarrow\Downarrow\rangle$-transition, while the green curve indicates the Ramsey fringes with the intermediate pulse. From the data, we see a phase offset of about $4\pi/5$ after application of the geometric phase gate, instead of the expected π (Fig. 5.4).

The origin of this discrepancy can partially be found in the presence of another, detuned, 2-level system: $|\uparrow\rangle$-$|\uparrow\downarrow\Uparrow\rangle$ (the other branches of the Λ-system are suppressed in view of the polarization selection rules: the laser field, which is aligned with the $|\downarrow\rangle$-$|\uparrow\downarrow\Downarrow\rangle$- ($|\uparrow\rangle$-$|\uparrow\downarrow\Uparrow\rangle$-) transition, is therefore anti-aligned with the other transitions of the Λ-system – we refer to Fig. 3.2 for the selection rules in the Voigt geometry). This other 2-level system undergoes a phase shift which is determined by the AC-Stark shift. For the parameters used in the experiment ($\Delta/\Omega_P \sim 5$), the counter-rotation can be found as follows ($t_{cycle} = 2\pi/|\Omega_P|$):

$$\begin{aligned}\phi_{net} &= \frac{|\Omega_P|^2}{4\Delta} \times t_{cycle} \\ &= \pi \times \frac{|\Omega_P|}{2\Delta} \sim \frac{\pi}{10}\end{aligned} \quad (5.11)$$

In addition, the nuclear spin effects described in Sect. 3.3.1 can give rise to additional, Ramsey-delay-dependent Overhauser shifts. From a numerical evaluation of Eq. 5.4, a detuning of only $\Omega_P/10 \sim 1$ GHz would be sufficient to generate the necessary additional phase shift. Such a detuning is very much within the reach of the Overhauser shift, and can therefore likely contribute to the offset.

5.4 Geometric Phases for 2-Qubit Entangling Gates

Pulsed, cyclic transitions resulting in geometric phases can be used to tweak the Larmor precession frequency: by adjusting the detuning and power or duration, a particular phase can be added to the Larmor precession, which could e.g. be used to synchronize different spin qubits with different Larmor precession frequencies.

Besides acting as a tuning knob for the effective Larmor frequency, geometric phases can be used to realize entangling, 2-qubit gates [1, 2]. Suppose we have a physical system where only one of four possible 2-qubit states is coupled to an excited state. This could e.g. occur due to different selection rules for electron singlets and trions, due to different detunings [1], etc. In such a system, a cyclic transition will add a π phase to only one of the states, say, $|\downarrow\rangle \otimes |\downarrow\rangle$. The effect of such a gate on all states is then as follows:

$$\begin{aligned}|\downarrow\rangle \otimes |\downarrow\rangle &\rightarrow -|\downarrow\rangle \otimes |\downarrow\rangle \\ |\downarrow\rangle \otimes |\uparrow\rangle &\rightarrow |\downarrow\rangle \otimes |\uparrow\rangle \\ |\uparrow\rangle \otimes |\downarrow\rangle &\rightarrow |\uparrow\rangle \otimes |\downarrow\rangle \\ |\uparrow\rangle \otimes |\uparrow\rangle &\rightarrow |\uparrow\rangle \otimes |\uparrow\rangle\end{aligned} \quad (5.12)$$

Such a gate is known as a controlled-Z-gate (CZ). By combining it with a Hadamard transformation of the target-qubit before and after the CZ-gate, we have that:

$$|\uparrow\rangle \otimes |\uparrow\rangle \rightarrow |\uparrow\rangle \otimes \frac{1}{\sqrt{2}}(|\uparrow\rangle + |\downarrow\rangle) \rightarrow |\uparrow\rangle \otimes \frac{1}{\sqrt{2}}(|\uparrow\rangle + |\downarrow\rangle) \rightarrow |\uparrow\rangle \otimes |\uparrow\rangle$$

$$|\uparrow\rangle \otimes |\downarrow\rangle \rightarrow |\uparrow\rangle \otimes \frac{1}{\sqrt{2}}(|\uparrow\rangle - |\downarrow\rangle) \rightarrow |\uparrow\rangle \otimes \frac{1}{\sqrt{2}}(|\uparrow\rangle - |\downarrow\rangle) \rightarrow |\uparrow\rangle \otimes |\downarrow\rangle$$

$$|\downarrow\rangle \otimes |\uparrow\rangle \rightarrow |\downarrow\rangle \otimes \frac{1}{\sqrt{2}}(|\uparrow\rangle + |\downarrow\rangle) \rightarrow |\downarrow\rangle \otimes \frac{1}{\sqrt{2}}(|\uparrow\rangle - |\downarrow\rangle) \rightarrow |\downarrow\rangle \otimes |\downarrow\rangle$$

$$|\downarrow\rangle \otimes |\downarrow\rangle \rightarrow |\downarrow\rangle \otimes \frac{1}{\sqrt{2}}(|\uparrow\rangle - |\downarrow\rangle) \rightarrow |\downarrow\rangle \otimes \frac{1}{\sqrt{2}}(|\uparrow\rangle + |\downarrow\rangle) \rightarrow |\downarrow\rangle \otimes |\uparrow\rangle \quad (5.13)$$

When we compare Eq. 5.13 with Eq. 1.21, we see that the CZ-gate and the CNOT gate are equivalent to within a basis transformation, and that the CZ-gate is therefore an entangling gate.[1] Physical implementations of such geometric phases for use in 2-qubit gates were proposed and demonstrated for charged, self-assembled quantum dots in Refs. [2] and [1].

References

1. D. Kim, S. G. Carter, A. Greilich, A. S. Bracker, and D. Gammon. Ultrafast optical control of entanglement between two quantum-dot spins. *Nat. Phys.*, 7:223, 2011.
2. T. D. Ladd and Y. Yamamoto. Simple quantum logic gate with quantum dot cavity QED systems. *Phys. Rev. B*, 84:235307, 2011.
3. A. Messiah. *Quantum mechanics*. Dover, 1999.
4. S. E. Economou, L. J. Sham, Y. Wu, and D. G. Steel. Proposal for optical U(1) rotations of electron spin trapped in a quantum dot. *Phys. Rev. B*, 74:205415, 2006.
5. S. E. Economou and T. L. Reinecke. Theory of Fast Optical Spin Rotation in a Quantum Dot Based on Geometric Phases and Trapped States. *Phys. Rev. Lett.*, 99:217401, 2007.
6. D. Press, K. De Greve, P. McMahon, T. D. Ladd, B. Friess, C. Schneider, M. Kamp, S. Hfling, A. Forchel, and Y. Yamamoto. Ultrafast optical spin echo in a single quantum dot. *Nat. Photonics*, 4:367, 2010.
7. E. Kim, K. Truex, X. Xu, B. Sun, D. G. Steel, A. S. Bracker, D. Gammon, and L. J. Sham. Fast spin rotations by optically controlled geometric phases in a charge-tunable inas quantum dot. *Phys. Rev. Lett.*, 104(16):167401, 2010.
8. M. A. Nielsen and I. L. Chuang. *Quantum Computation and Quantum Information*. Cambridge University Press, 2000.

---

[1] The same argument is valid for any conditional phase gate, where only one 2-qubit combination undergoes a geometric phase shift of π – we refer to Ref. [8] for more details.

Chapter 6
Ultrafast Optical Control of Hole Spin Qubits: Suppressed Nuclear Feedback Effects

The strong contact hyperfine interaction between a single electron and the nuclear spins in a quantum dot [1–5] represents a severe limitation on its use as a qubit, as we discussed in Sect. 3.3.1. In particular, it results in non-linear, non-Markovian dynamics of the electron-nuclear system [6–9], making the use of the Larmor-precession as a single qubit gate cumbersome. While dynamical decoupling, in essence an extension of the simple spin echo presented in Sect. 3.3.2 to multiple refocusing pulses, could be used to decouple the effects of the nuclear spin bath [10–14], avoiding the contact hyperfine interaction altogether would be a much simpler solution. For some materials, this could be achieved by careful isotopic engineering: examples for group-IV and II–VI materials can be found in Refs. [15–19]. However, In, Ga and As do not possess any spin-0 isotopes, making this approach impossible for the self-assembled quantum dots used in this work.

The different wavefunctions of electrons and holes provide an alternative way of overcoming interactions with the nuclear spins. Figure 6.1 provides a schematic overview: as we derived in Sect. 2.1, the underlying band-wavefunction of electrons has s-type symmetry, while that of holes has p-type symmetry. It is this p-type symmetry that causes the hole to not have any probability amplitude at the core of the nuclear spins, resulting in a reduced hyperfine interaction [20–23], dominated by dipolar terms.

We inadvertently already described the hyperfine interaction of a single quantum dot hole state in Sect. 3.3.1, in the context of nuclear feedback effects for *electron* spins: as we showed there, the unpaired hole of an electron trion has a profoundly different hyperfine interaction than the contact hyperfine interaction of a single electron spin. Summarizing that treatment, an unpaired hole in a single quantum dot has predominantly heavy-hole (HH) character due to the quantization along the growth axis, with a small light-hole (LH) component mixed in (we use the same convention as before, where the growth axis is along the z-direction):

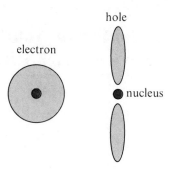

Fig. 6.1 Schematic overview of the difference of electrons and holes in terms of overlap with the nuclear spins. The *yellow shaded area* represents the band-wavefunction for electrons (*left*, s-type) and holes (*right*, p-type symmetry), while the *red circles* indicate the location of the nuclei. Note the absence of any overlap between the band-wavefunction and the nucleus for holes, which explains the lack of contact hyperfine interaction

$$|\Uparrow\rangle_z \equiv \frac{1}{\sqrt{1+|\eta|^2}}\left[|J_z = 3/2\rangle + \eta^+ |J_z = -1/2\rangle\right] \quad (6.1)$$

$$|\Downarrow\rangle_z \equiv \frac{1}{\sqrt{1+|\eta|^2}}\left[|J_z = -3/2\rangle + \eta^- |J_z = 1/2\rangle\right]. \quad (6.2)$$

In Eq. 6.2, $\eta^\pm = |\eta|e^{\pm i\xi}$, where ξ refers to a preferential axis of high symmetry. The heavy-hole light-hole mixing can be caused by any effect that breaks the cylindrical symmetry – in practice, strain and quantum dot asymmetry. For typical dots, $|\eta|$ is around 0.2; the amount of HH-LH mixing can be experimentally determined by studying the optical selection rules – we refer to Appendix C for additional details. The confinement energy in the quantum dot typically separate these two (quasi-) heavy-hole states from any other, such that this heavy-hole manifold and 2-level system can be considered as a pseudo-spin-1/2 system.

To lowest order, a magnetic field in the Voigt geometry (say, along the x-direction) will hybridize the heavy-hole states:

$$|\Uparrow\rangle_x = \frac{1}{\sqrt{2}}(|\Uparrow\rangle_z + |\Downarrow\rangle_z) \quad (6.3)$$

$$|\Downarrow\rangle_x = \frac{1}{\sqrt{2}}(|\Uparrow\rangle_z - |\Downarrow\rangle_z) \quad (6.4)$$

Transforming into a pseudo-spin-1/2 description [20], where $|\Uparrow, \Downarrow\rangle_z = |S_z = \pm 1/2\rangle$ and $|\Uparrow, \Downarrow\rangle_x = |S_x = \pm 1/2\rangle$, the dipolar hyperfine Hamiltonian can be written in the following way [8]:

$$H_{\text{dipole}} \sim \sum_{i=\text{nuclei}} A_{\text{dipole},i}\left[I_{z,i}S_z + O(\eta)(I_{x,i}S_x + I_{y,i}S_y) + O(\eta^2)(I_{z,i}S_x + I_{z,i}S_y)\right]$$
$$(6.5)$$

The amplitude, $A_{\text{dipole},i}$ of the dipolar hyperfine interaction is typically some 10–20% of the contact hyperfine Hamiltonian, and has been measured experimentally for hole-charged quantum dots [24, 25]. This reduction in hyperfine amplitude, however, is not the only advantage of hole spins. The leading term in Eq. 6.5, $I_{z,i}S_z$, acts as an effective magnetic field in the z-direction, perpendicular to the external magnetic field, and therefore does not contribute to dephasing or decoherence[1]: this Ising Hamiltonian therefore further suppresses dephasing and decoherence due to nuclear spins. It is only the second and third terms, of order $O(\eta)$ and $O(\eta^2)$, that contribute to variations in the effective magnetic field along the x-direction. The combined effect strongly depends on the amount of LH-inmixing, but one could theoretically realize a suppression by several orders of magnitude of the nuclear hyperfine interaction.

While direct microwave manipulation of the hole states would in principle be possible, it was shown in Ref. [27] that the coupling of the heavy holes to a microwave field is orders of magnitude weaker than for electrons, making such manipulation quite impractical. The p-type symmetry of the hole-wavefunction does allow for different spin-orbit coupling mechanisms to couple to the pseudo-spins, which could in theory be exploited for all-electrical manipulation (see e.g. Refs. [28, 29] for examples of electric-field, spin-orbit based spin manipulation techniques), but none have been implemented so far. All-optical manipulation techniques, however, can be applied, and were previously used for fast initialization of a single hole [22] and coherent population trapping [23].

The lowest-energy, optically excited states are hole-trions, consisting of an antisymmetric (singlet) combination of heavy-holes, and an unpaired electron spin, as indicated in Fig. 6.2a. The optical selection rules can now be derived in a manner very similar to the methods used in Sect. 2.3, and are summarized in Fig. 6.2c (we assume a small amount of LH-inmixing only, such that the zeroth-order selection rules for pure heavy holes are approximately valid). This level structure is very much equivalent to that of a single electron spin in Voigt geometry: here as well, two Λ-systems are present, that couple both of the heavy-hole ground states to each of the trions. Therefore, one would expect that the same optical manipulation techniques as for electron spins can be applied to hole spins, such as the ultrafast control techniques described in Chap. 3. In the remainder of this chapter, we will describe the first realization of a full hole-(pseudo-)spin qubit, which we previously reported in Ref. [30].

---

[1]This is true to first order: this term can give rise to spin flips in the x-direction (we can rewrite it in terms of lowering and raising operators along x), though these will be energetically forbidden and therefore only contribute to higher order. We refer to any of the theoretical descriptions of the full central-spin problem for details, e.g. Refs. [1, 4, 26].

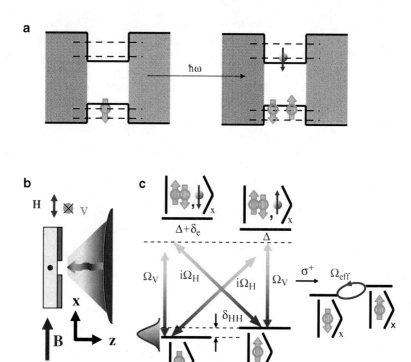

Fig. 6.2 Schematic overview of the all-optical control of a single hole spin. (**a**) A single hole is manipulated by (virtual) excitation of a hole-trion; (**b**) axis convention used, with a magnetic field in Voigt geometry; (**c**) the relevant energy levels, and the selection rules. In a similar way as was derived in Sect. 3.1.1.1, the effect of a detuned, circular lase pulse can be summarized by an effective Rabi-oscillation around an axis along the growth-direction, z

6.1 Device Design

In addition to carbon-δ-doped devices, similar to the electron-doped ones used in Chap. 3, we designed a new type of device that allows for deterministic charging of a quantum dot with a single hole ('charge-tuneable' device). By embedding the quantum dots in an asymmetric p-i-n diode, close to a p-doped region, a single hole can tunnel into the quantum dot at a particular voltage bias of the diode. Such a structure was previously reported for both electron and hole-charged quantum dots (see Refs. [22, 23, 31]), and allows for unambiguous determination of the charge state based on the characteristic charging energies upon changing the voltage bias.[2] Both designs contain an asymmetric optical cavity that preferentially directs

---

[2] As opposed to the δ-doping devices, where it would in principle be possible for a single, spurious electron to be present inside the quantum dot, which cannot be distinguished from a single hole as the optical signatures are the same.

6.2 Ultrafast Coherent Control

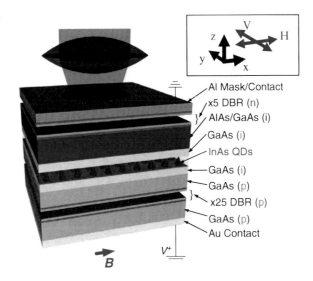

Fig. 6.3 Schematic overview of the device structure used for deterministic hole-charging of a single quantum dot. See text for details (Figure reproduced from Ref. [30])

the photoluminescence towards the collection optics, and the low dot density, in combination with spectral inhomogeneity, guarantees that only a single hole is studied. For the remainder, we focus on the results obtained with the charge-tuneable device, which is schematically illustrated in Fig. 6.3. We refer to Appendix E for further details on the device designs used.

The magneto-photoluminescence of the charge-tuneable devices is shown in Fig. 6.4a. By comparing the energetic differences between the charge-states of the respective voltage-plateaus [31], we can identify the regime where a single hole is stably trapped inside the quantum dot: the X^+-region (we refer to Appendix E for further details). Figure 6.4b displays the polarization-resolved photoluminescence, showing polarization selection rules that are equivalent to the ones shown in Fig. 6.2c: two optical Λ-systems are indeed present, permitting all-optical, ultrafast control of the hole spin ground states.

6.2 Ultrafast Coherent Control

The experimental setup used is identical to the one shown in Fig. 3.7, with the addition of voltage control of the quantum dot sample (low-frequency bias lines are mounted inside the superconducting cryostat and used for that purpose). A magnetic field of 8 T is used, in Voigt geometry, resulting in an effective Larmor frequency ($\delta_{HH}/2\pi$) for the heavy-hole pseudo-spin of 30.2 GHz. Before applying any rotation pulses, the system is initialized into the $|\Uparrow\rangle_x$-state by means of a narrowband, continuous-wave (CW) optical pumping laser resonant with the $|\Downarrow\rangle_x$-$|\Uparrow\Downarrow\downarrow\rangle_x$-transition. As for the electron spin, this optical pumping laser again also acts

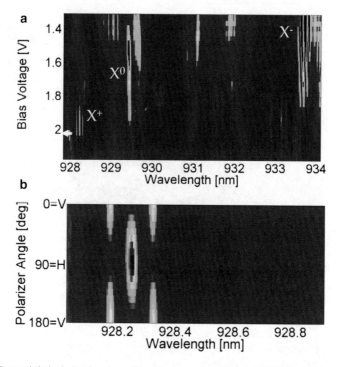

Fig. 6.4 Deterministic hole-charging of a single quantum dot. (**a**) Magneto-PL of a single quantum dot, for different electrical bias conditions. Comparison with theoretical spectra allows for determination of the charge-states. (**b**) Polarization-resolved magneto-PL for the hole-charged quantum dot in (**a**): note the optical selection rules, in good agreement with Fig. 6.2c (Figures reproduced from Ref. [30])

as the spin readout laser: upon excitation to the $|\Downarrow\rangle_x$-state, a single, cross-polarized photon will be emitted along the $|\Uparrow\rangle_x$-$|\Uparrow\Downarrow\downarrow\rangle_x$-transition, which can be spectrally and polarization-filtered and detected on a single photon counter.

The ultrafast rotations are applied using 3–4 ps-, circularly polarized pulses from a modelocked laser, detuned by about 340 GHz from the hole-trion transitions. The entire optical manipulation scheme is illustrated in Fig. 6.5.

The effect of a single rotation pulse is illustrated in Fig. 6.6: when continuously varying the power of the rotation pulse used, Rabi-oscillations can be observed, as was the case for electron spins. In Appendix D, these Rabi-oscillations are modelled in detail based on an AC-Stark description (see also Sect. 3.1.2). However, as we showed before, a description in terms of stimulated-Raman-transitions would be equivalent [32, 33].

In addition to the ultrafast optical spin rotations (rotation around the z-axis), the Larmor precession of the hole pseudo-spin can be used for rotation around the x-axis. This can be visualized using Ramsey interferometry, as shown in Fig. 6.7.

6.2 Ultrafast Coherent Control

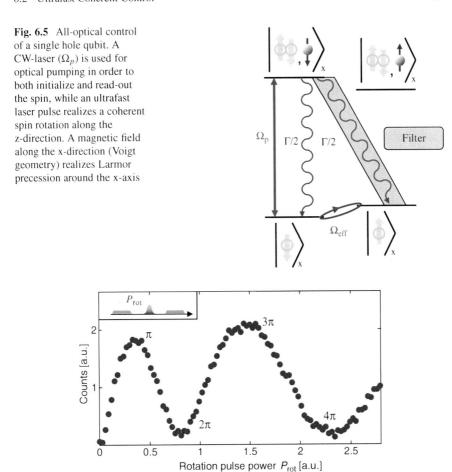

Fig. 6.5 All-optical control of a single hole qubit. A CW-laser (Ω_p) is used for optical pumping in order to both initialize and read-out the spin, while an ultrafast laser pulse realizes a coherent spin rotation along the z-direction. A magnetic field along the x-direction (Voigt geometry) realizes Larmor precession around the x-axis

Fig. 6.6 Rabi-oscillations of a single hole qubit. *Inset*: pulse scheme used (Figure reproduced from Ref. [30])

In addition, Ramsey interferometry allows for extraction of the fidelity of a single ultrafast hole rotation – we refer to Appendix A.2, in which we extracted fidelities upwards of 90–95 % for the experiments in Figs. 6.6 and 6.7.

Combining ultrafast, pulse-induced control with Larmor-precession now realizes full SU(2) control over a hole (pseudo-) spin qubit, as any coherent control operation can be decomposed into a series of rotation pulses with intermittent Larmor precession [34]. Figure 6.8 illustrates how the entire surface of the Bloch sphere can be traced out using these two control operations, with the time needed for a single qubit operation at 20 ps or less.

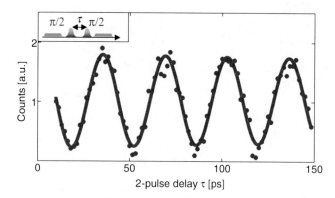

Fig. 6.7 Ramsey-interference of a single hole qubit. *Inset*: pulse scheme used (Figure reproduced from Ref. [30])

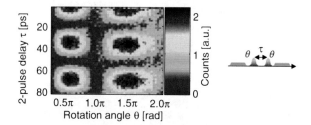

Fig. 6.8 Full SU(2) control of a single hole qubit. By varying both the rotation pulse power (Rabi-angle, θ) and the Larmor precession duration, τ, the entire surface of the Bloch sphere can be traced out. *Inset*: pulse scheme used (Figure reproduced from Ref. [30])

6.3 Suppressed Nuclear Feedback Effects

While the ultrafast optical control techniques described above allow for complete SU(2) control of a hole qubit, they can also be used to study the effects of the nuclear hyperfine interaction on the dynamics of hole spins. Theoretically, one would expect a significant suppression of hyperfine effects for holes, in view of the absence of a contact hyperfine interaction, and the particular nature of the dipolar hyperfine coupling in the Voigt-geometry (see Eq. 6.5).

6.3.1 Absence of Nuclear-Induced Hysteresis

For electron spins, the effects of the contact hyperfine Hamiltonian on the spin dynamics were studied in Sect. 3.3.1, and analyzed in detail in Appendices B.1 and B.2. The dynamics are pronouncedly non-Markovian, due to the development of strong feedback loop between the electron spin dynamics and the nuclear spins.

6.3 Suppressed Nuclear Feedback Effects

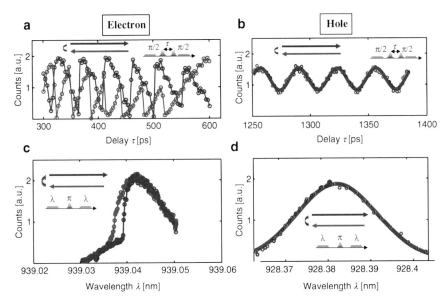

Fig. 6.9 Re-emergence of hysteresis-free dynamics for hole spins (Figures reproduced from [30]) (**a**) Asymmetric and hysteretic Ramsey fringes for an electron spin in a QD. The *green* and *blue circles* refer respectively to backwards and forwards scanning of the delay line, as indicated by the *arrows* [9]. *Inset*: the pulse sequence used in the experiment. (**b**) Symmetric and hysteresis-free Ramsey fringes for a hole spin in a QD. Even for large delays, the Ramsey fringes remain perfectly sinusoidal. The *green* and *blue circles* refer to different scanning directions, as indicated by the *arrows*. The *red curve* represents a sinusoidal least-squares fit. *Inset*: pulse sequence used. (**c**) Effective absorption signal for an electron spin in a QD, showing clear asymmetry and hysteresis upon scanning the pump laser wavelength in different directions (*green* and *blue circles*, as indicated by the *arrows*). *Inset*: pulse sequence used. (**d**) Effective absorption signal for a hole spin in a QD. No hysteresis or asymmetry was observed upon scanning in different directions (*green* and *blue circles*, as indicated by the *arrows*). *Red curve*: least-squares fit of a Gaussian absorption profile. *Inset*: pulse sequence used

The feedback loop develops a large nuclear spin polarization (dynamic nuclear polarization), which in turn couples back to the electron spin through the Overhauser shift. This non-Markovian behavior reflects itself in asymmetric, hysteretic Ramsey-interference fringes (Fig. 6.9a), and asymmetric, hysteretic line shapes upon scanning a laser through an effective resonant-absorption transition (see Appendix B.2 and Fig. 6.9). As the effective magnetic field is used, via the Larmor precession, as a control field for the electron spin, this non-Markovian dynamics strongly affects the controllability of electron spins.

In Fig. 6.9a, b, we compare the Ramsey interference fringes for a single electron spin and a hole pseudo-spin. For a single electron spin, we observe a pronounced hysteresis, and strong asymmetry. For a HH qubit, however, and as shown in Fig. 6.9b, no such hysteresis is observed. This lack of observable hysteresis is understood to be due to the strong reduction of Overhauser shifts that results from the absence of a contact hyperfine coupling for holes. We observe a reduction by least a factor of 30 in the nuclear feedback strength when using holes instead of electrons, in line with recent direct measurements of the reduced hyperfine coupling of holes [24,25,35,36]. This estimate is a conservative lower bound, limited by hole decoherence that may obscure possible weak effects – we refer to Appendix B.3 for further details.

Similarly, we can compare effective resonant-absorption experiments for electrons and holes, as shown in Fig. 6.9c, d. In order to avoid optical pumping, a π-rotation pulse is applied in between each initialization cycle (inset of Fig. 6.9c, d). This approach mimics two-CW-laser resonant absorption experiments performed for electrons [7, 8] and for holes [23, 24]. For electron spins scanning the laser through the $|\downarrow\rangle$-$|\downarrow\uparrow,\Uparrow\rangle$-resonance leads to a hysteretic nuclei-induced wandering of that resonance; this effect is seen in Fig. 6.9c and is analyzed in detail in Appendix B.2. In contrast, no hysteresis is observed for a hole-charged QD. The absorption profile is completely symmetric, and is best fit by a Gaussian (red curve in Fig. 6.9d) with a linewidth of 6.7 GHz. The notable broadening does suggest a more significant spectral diffusion of hole-charged QDs, as opposed to electron-doped ones – similar to the effects observed in Ref. [23].

Together, these studies demonstrate that the hole qubit is superior to the electron in terms of interactions with the nuclear spins, and by extension, in terms of controllability by means of the Larmor-precession frequency.

6.3.2 T_2^* and Electrical Noise

Despite the reduction in nuclear feedback strength, a study of the T_2^* decoherence as shown in Fig. 6.10a, b yields very similar results as for a single electron spin (see Sect. 3.3.2, where we used an echo technique to extract an electron T_2^* of about 1.5 ns): using Ramsey-interferometry, and untainted by nuclear feedback effects, we find that the hole Ramsey fringes in Fig. 6.10a decay with a Gaussian envelope function (i.e. $\propto \exp[-(t/T_2^*)^2]$), characterized by a T_2^*-time constant of 2.3 ns (Fig. 6.10b). This value is independent of the magnetic field for fields between 6 and 10 T. Given the observed reduction in nuclear feedback strength observed before, and in view of theoretical predictions of strongly suppressed nuclear hyperfine interactions [21, 37], it is unlikely that nuclear spin effects are responsible for the T_2^* times we observe. Non-nuclear dephasing processes are far more likely; such processes limit coherence in localized hole spins in quantum wells [38, 39], and may even overwhelm nuclear processes in electron-charged quantum dots at some magnetic fields, depending on the dot's proximity to noisy surface states [5]. These dephasing processes generally arise from fluctuations in the localizing potential, a process that may be observed as spectral diffusion of the optical transitions, evidenced by the increased optical linewidth in the resonance-scanning experiment described above.

The effect of spectral diffusion on hole spin coherence, in the form of randomly varying electric fields, can be directly examined using our charge-tunable devices.

6.3 Suppressed Nuclear Feedback Effects

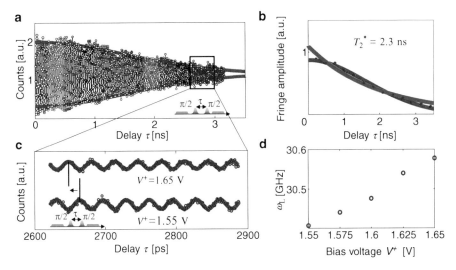

Fig. 6.10 T_2^* and electrical dephasing (Figures reproduced from [30]). (**a**) Ramsey fringes as a function of the delay τ between two $\pi/2$ pulses. The different colors refer to different positions of the stage on the rail (see text), while the *red envelopes* indicate a least-squares-fit Gaussian decay with $T_2^* = 2.3$ ns. *Inset*: pulse timing used. *Boxed area*: delays used in **c**. (**b**) Amplitude of the Ramsey fringes, as a function of the delay τ. *Red curve*: fit to a Gaussian decay ($T_2^* = 2.3$ ns); *green curve*: fit to an exponential decay. (**c**) Long-delay Ramsey fringes, for different applied voltage bias V^+. *Top*: $V^+ = 1.65$ V; *bottom*: $V^+ = 1.55$ V. *Red curves*: least-squares, sinusoidal fit. The cumulative effect of the difference in Larmor precession frequency ω_L can be seen in the shifting of the curves. (**d**) Larmor-precession frequency ω_L as a function of the externally applied bias voltage V^+

By deliberately changing the electrical bias over the QD, we can Stark-shift the quantum dot. By controlling the electrical bias such that the resulting spectral shifts are similar in magnitude to those presumed responsible for spectral diffusion, we measure notably different effective Larmor precession frequencies of the HH pseudo-spin. In Fig. 6.10c, the cumulative effect of such a difference in Larmor frequency is shown through long-delay Ramsey fringes. The different precession frequencies lead to anti-phase Ramsey fringes for delays similar to T_2^*. In Fig. 6.10d, the monotonic increase of the Larmor frequency with applied bias is shown. These results indicate that electrical fields couple strongly to the spin coherence of the hole qubits, and that T_2^* is actually limited by electric field fluctuations rather than nuclear hyperfine effects. In addition, similar values for T_2^* are obtained for the QDs in the δ-doped sample, suggesting that the actual source of the charge fluctuations is reasonably uniform from sample to sample, and not directly related to a particular device or quantum dot studied. The microscopic coupling mechanism to the hole spins is currently not well understood, but is assumed to be related to the strong spin-orbit coupling of hole spins.

Our measurement of T_2^* contrasts markedly with an estimate of T_2^* obtained via coherent population trapping in similar HH-doped QDs [23]. However, that experiment effectively filters out a hole precession process for one particular spectral location, by deconvolving the effects of spectral diffusion which are assumed to not couple to the spin coherence. As our studies indicated, this assumption is not valid for generic quantum dots. By removing the potential dephasing effects of spectral diffusion, substantially longer values of T_2^* can of course be inferred, yet it is highly doubtful that these correspond to the true dephasing time.

While the ability of electric fields to shift the effective QD Larmor frequency may unfortunately impact T_2^*, it simultaneously provides a convenient advantage of hole pseudo-spin qubits over electron-spin qubits: namely, it introduces the ability to locally tune multiple qubits into resonance, which may have important implications for viable 2-qubit gates, and aid in scalability to many-qubit systems.

6.4 Hole Spin Echo and T_2

Finally, we use a spin-echo technique [5] to measure the T_2-decoherence time of the HH qubit. Figure 6.11a illustrates the fringes obtained from scanning the π-echo-pulse, and Fig. 6.11b shows the fringe contrast as a function of total delay $2T$ used in the spin-echo Ramsey fringes. The decay is best fit by an exponential (i.e., $\propto \exp[-(t/T_2)]$), with $T_2 = 1.1\,\mu\text{s}$, the same order of magnitude as for electron spins [5]. Here as well, magnetic-field-dependent studies do not show any dependence on the field between 6 and 10 T. In combination with a single-qubit operation time of about 20 ps, this T_2-value predicts a figure of merit of about 50,000 operations within the coherence time of the qubit (operational fidelity permitting: see Sect. 3.3.2). Although 500 times longer than T_2^*, the obtained T_2-value is still lower than what is theoretically expected for a nuclear-spin-limited decay. Phonon interactions are expected to only weakly affect the quantum dot hole spin [20, 21, 40], and can therefore also be excluded as a dominant source of decoherence. It therefore appears that T_2 is limited by a similar, non-nuclear

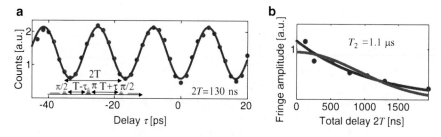

Fig. 6.11 All-optical spin echo for a single hole qubit. (**a**) Spin-echo Ramsey fringes as a function of the imbalance, τ, between two halves of the spin echo experiment. *Inset*: pulse timing used. (**b**) Amplitude of the spin-echo fringes, as a function of the delay $2T$ between the $\pi/2$ pulses. *Green curve*: fit to a Gaussian decay; *red curve*: fit to an exponential decay ($T_2 = 1.1\,\mu\text{s}$) (Figures reproduced from [30])

mechanism as that which most likely limits T_2^*, i.e. charge-induced spectral diffusion. Such a process can be mimicked by AC-modulation of the external voltage bias, and in initial experiments, we have indeed observed some suppression of T_2 when introducing such a modulation. We measured significant dot-to-dot variance of T_2, sometimes measuring T_2 as low as several hundred ns. The variation is likely due to differences in the spin-orbit contribution to the hole-pseudospin Hamiltonian, in large part due to different HH-LH mixing for dots of different shapes and levels of strain, and differences in the effective charge noise bath for different QDs. Further understanding of the decoherence mechanisms for holes may enable extension of the spin coherence through further device engineering, as well as through the application of advanced dynamical decoupling schemes, as have been demonstrated for electron spins [14].

References

1. W. A. Coish and D. Loss. Hyperfine interaction in a quantum dot: Non-Markovian electron spin dynamics. *Phys. Rev. B*, 70:195340, 2004.
2. J. R. Petta *et al*. Coherent manipulation of coupled electron spins in semiconductor quantum dots. *Science*, 309:2180, 2005.
3. W. Yao, R.-B. Liu, and L. J. Sham. Theory of electron spin decoherence by interacting nuclear spins in a quantum dot. *Phys. Rev. B*, 74:195301, 2006.
4. W. M. Witzel and S. Das Sarma. Quantum theory for electron spin decoherence induced by nuclear spin dynamics in semiconductor quantum computer architectures: Spectral diffusion of localized electron spins in the nuclear solid-state environment. *Phys. Rev. B*, 74:035322, 2006.
5. D. Press, K. De Greve, P. McMahon, T. D. Ladd, B. Friess, C. Schneider, M. Kamp, S. Höfling, A. Forchel, and Y. Yamamoto. Ultrafast optical spin echo in a single quantum dot. *Nat. Photonics*, 4:367, 2010.
6. I. T. Vink *et al*. Locking electron spins into magnetic resonance by electron–nuclear feedback. *Nat. Phys.*, 5:764–768, 2009.
7. C. Latta *et al*. Confluence of resonant laser excitation and bidirectional quantum-dot nuclear-spin polarization. *Nat. Phys.*, 5:758, 2009.
8. X. Xu *et al*. Optically controlled locking of the nuclear field via coherent dark-state spectroscopy. *Nature*, 459(4):1105, 2009.
9. T. D. Ladd, D. Press, K. De Greve, P. McMahon, B. Friess, C. Schneider, M. Kamp, S. Höfling, A. Forchel, and Y. Yamamoto. Pulsed nuclear pumping and spin diffusion in a single charged quantum dot. *Phys. Rev. Lett.*, 105:107401, 2010.
10. L. Viola and S. Lloyd. Dynamical suppression of decoherence in two-state quantum systems. *Phys. Rev. A*, 58:2733–2744, 1998.
11. M.J. Biercuk, H. Uys, A. P. VanDevender, N. Shiga, W. M. Itano, and J. J. Bollinger. Optimized dynamical decoupling in a model quantum memory. *Nature*, 458:996, 2009.
12. J. Du, X. Rong, N. Zhao, Y. Wang, J. Yang, and R. B. Liu. Preserving electron spin coherence in solids by optimal dynamical decoupling. *Nature*, 461:1265, 2009.
13. G. de Lange, Z. H. Wang, D. Ristè, V. V. Dobrovitski, and R. Hanson. Universal dynamical decoupling of a single solid-state spin from a spin bath. *Science*, 330(6000):60, 2010.
14. H. Bluhm, S. Foletti, I. Neder, M. Rudner, D. Mahalu, V. Umansky, and A. Yacoby. Dephasing time of GaAs electron-spin qubits coupled to a nuclear bath exceeding 200 μs. *Nat. Phys.*, 7:109, 2010.

15. T. D. Ladd, F. Jelezko, R. Laflamme, Y. Nakamura, C. Monroe, and J. L. O'Brien. Quantum computers. *Nature*, 464:45, 2010.
16. G. Balasubramanian et al. Ultralong spin coherence time in isotopically engineered diamond. *Nat. Mater.*, 8:383, 2008.
17. B. Kane. A silicon-based nuclear spin quantum computer. *Nature*, 393:133–137, 1998.
18. M. G. Borselli, R. S. Ross, A. A. Kiselev, E. T. Croke, K. S. Holabird, P. W. Deelman, L. D. Warren, I. Alvarado-Rodriguez, I. Milosavljevic, F. C. Ku, W. S. Wong, A. E. Schmitz, M. Sokolich, M. F. Gyure, and A. T. Hunter. Measurement of valley splitting in high-symmetry Si/SiGe quantum dots. *Appl. Phys. Lett.*, 98:123118, 2011.
19. K. De Greve, S. M. Clark, D. Sleiter, K. Sanaka, T. D. Ladd, M. Panfilova, A. Pawlis, K. Lischka, and Y. Yamamoto. Photon antibunching and magnetospectroscopy of a single fluorine donor in ZnSe. *Appl. Phys. Lett.*, 97:241913, 2010.
20. D. V. Bulaev and D. Loss. Spin decoherence and relaxation of holes in a quantum dot. *Phys. Rev. Lett.*, 95:076805, 2005.
21. J. Fischer, W. A. Coish, D. V. Bulaev, and D. Loss. Spin decoherence of a heavy hole coupled to nuclear spins in a quantum dot. *Phys. Rev. B*, 78:155329, 2008.
22. B. D. Gerardot et al. Optical pumping of a single hole spin in a quantum dot. *Nature*, 451:441, 2008.
23. D. Brunner, B. D. Gerardot, P. A. Dalgarno, G. Wst, K. Karrai, N. G. Stoltz, P. M. Petroff, and R. J. Warburton. A coherent single-hole spin in a semiconductor. *Science*, 325(5936):70–72, 2009.
24. P. Fellahi, S. T. Yilmaz, and A. Imamoglu. Measurement of a Heavy-Hole Hyperfine Interaction in InGaAs Quantum Dots Using Resonance Fluorescence. *Phys. Rev. Lett.*, 105:257402, 2010.
25. E. A. Chekhovich, A. B. Krysa, M. S. Skolnick, and A. I. Tartakovskii. Direct measurement of the hole-nuclear spin interaction in single quantum dots. *Phys. Rev. Lett.*, 106:027402, 2010.
26. R.-B. Liu, W. Yao, and L. J. Sham. Control of electron spin decoherence caused by electron–nuclear spin dynamics in a quantum dot. *New J. Phys.*, 9:226, 2007.
27. D. Sleiter and W. F. Brinkman. Using holes in GaAs as qubits: An estimate of the Rabi frequency in the presence of an external rf field. *Phys. Rev. B*, 74:153312, 2006.
28. K. C. Nowack, F. H. L. Koppens, Yu. V. Nazarov, and L. M. K. Vandersypen. Coherent Control of a Single Electron Spin with Electric Fields. *Science*, 318:1430, 2007.
29. S. Nadj-Perge, S. M. Frolov, E. P. A. M. Bakkers, and L. P. Kouwenhoven. Spin–orbit qubit in a semiconductor nanowire. *Nature*, 468:1084–1087, 2010.
30. K. De Greve, P. L. McMahon, D. Press, T. D. Ladd, D. Bisping, C. Schneider, M. Kamp, L. Worschech, S. Höfling, A. Forchel, and Y. Yamamoto. Ultrafast coherent control and suppressed nuclear feedback of a single quantum dot hole qubit. *Nat. Phys.*, 7:872, 2011.
31. R. J. Warburton, C. Schäflein, D. Haft, F. Bickel, A. Lorke, K. Karrai, J. M. Garcia, W Schoenfeld, and P. M. Petroff. Optical emission from a charge-tunable quantum ring. *Nature*, 405:926, 2000.
32. S. E. Economou, L. J. Sham, Y. Wu, and D. G. Steel. Proposal for optical U(1) rotations of electron spin trapped in a quantum dot. *Phys. Rev. B*, 74:205415, 2006.
33. S. M. Clark, K-M. C. Fu, T. D. Ladd, and Y. Yamamoto. Quantum computers based on electron spins controlled by ultrafast off-resonant single optical pulses. *Phys. Rev. Lett.*, 99:040501, 2007.
34. M. A. Nielsen and I. L. Chuang. *Quantum Computation and Quantum Information*. Cambridge University Press, 2000.
35. B. Eble, C. Testelin, P. Desfonds, F. Bernardot, A. Balocchi, T. Amand, A. Miard, A. Lemaitre, X. Marie, and M. Chamarro. Hole–Nuclear spin interaction in quantum dots. *Phys. Rev. Lett.*, 102:146601, 2009.
36. C. Testelin, F. Bernardot, B. Eble, and M. Chamarro. Hole–spin dephasing time associated with hyperfine interaction in quantum dots. *Phys. Rev. B*, 79:195440, 2009.
37. J. Fischer and D. Loss. Hybridization and Spin Decoherence in Heavy-Hole Quantum Dots. *Phys. Rev. Lett.*, 105:266603, 2010.

38. Y. G. Semenov, K. N. Borysenko, and K. W. Kim. Spin-phase relaxation of two-dimensional holes localized in a fluctuating potential. *Phys. Rev. B*, 66:113302, 2002.
39. M. Syperek, D. R. Yakovlev, A. Greilich, J. Misiewicz, M. Bayer, D. Reuter, and A. D. Wieck. Spin coherence of holes in GaAs/(Al,Ga)As quantum wells. *Phys. Rev. Lett.*, 99:187401, 2007.
40. D. Heiss, S. Schaeck, H. Huebl, M. Bichler, G. Abstreiter, J. J. Finley, D. V. Bulaev, and D. Loss. Observation of extremely slow hole spin relaxation in self-assembled quantum dots. *Phys. Rev. B*, 76:241306(R), 2007.

Chapter 7
Entanglement Between a Single Quantum Dot Spin and a Single Photon

While fast, complete SU(2) control of a long-lived qubit is an essential ingredient of any quantum information system, our focus on quantum communication systems, and quantum repeaters in particular, requires a means of interfacing the stationary, matter and memory qubits, with flying, photonic qubits. As we derived in Sects. 1.1.3.1 and 1.4, entanglement between a spin-qubit and a photonic qubit [1–5] provides exactly the type of interface required in order to generate remotely-entangled spin-pairs and implement the quantum teleportation and entanglement swapping schemes that are at the heart of quantum repeaters [6–11].

In addition, when combined with a photonic qubit that is compatible with low-loss optical fiber technology (i.e., at telecom wavelengths, such as the lowest-loss, 1.5 μm transmission band of commercial fiber [12]), such entanglement may be distributed over long distances. Self-assembled quantum dots, in view of their fast radiative recombination rates or, equivalently, strong optical dipole moments [13, 14], would form an ideal platform for such a quantum interface. Moreover, the III–V optical quantum dots are perfectly compatible with state-of-the art fabrication of high-quality optical microcavities, which could further improve the photon extraction efficiency and increase the overall photon yield [15].

In this chapter, we will describe a proof-of-principle experiment, where we develop an ultrafast frequency downconversion technique to 1,560 nm in order to measure spin-photon entanglement at near-infrared frequencies (NIR, 910 nm), which can also be used for generating a full 1,560 nm qubit, entangled with the quantum dot spin.

7.1 Spin-Photon Entanglement: Λ-System Decay

For a charged quantum dot, in the Voigt geometry, the optical selection rules present a double Λ-system, which we previously exploited for ultrafast manipulation of both electron and hole spins (see Chaps. 3 and 6). However, upon excitation of one of the

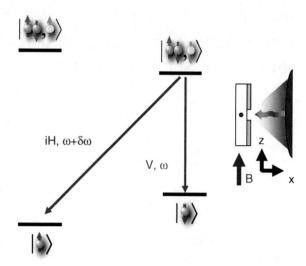

Fig. 7.1 Entanglement from Λ-system decay. Interference between the different decay-pathways in an optically active Λ-system results in spin-photon entanglement. *Inset*: Voigt-geometry used to realize a Λ-system configuration for charged quantum dots (Figures adapted from [16])

excited trion states, interference between the two possible pathways in the Λ-system results automatically in spin-photon entanglement, as was shown previously for single ions [4], atoms [2] and nitrogen-defect centers in diamond [3] (see Fig. 7.1):

$$|\Psi_{\text{entangled}}\rangle = \frac{1}{\sqrt{2}}(i|\uparrow\rangle \otimes |H\rangle \otimes |\omega + \delta\omega\rangle + |\downarrow\rangle \otimes |V\rangle \otimes |\omega\rangle) \quad (7.1)$$

In Eq. 7.1, H and V represent horizontal (parallel to the magnetic field) and vertical polarization states respectively, while $\delta\omega$ stands for the difference in frequency of the respective frequency components. This state is entangled in polarization and frequency, which can be verified by measuring in different bases of the spin and the photon, as argued in Sect. 1.1.3. However, the fact that the spin-photon entanglement is both in frequency and polarization, rather than in one degree of freedom only, makes verification non-trivial, as which-path information from e.g. the frequency can obscure a measurement of the polarization-entanglement, and vice-versa.

7.1.1 Measurement of Polarization-Entanglement: Frequency-Which-Path Information

We can reformulate the entangled state in Eq. 7.1 by moving to a density matrix description:

$$\rho_{\text{entangled}} = \frac{1}{2}(i|\uparrow\rangle \otimes |H\rangle \otimes |\omega + \delta\omega\rangle + |\downarrow\rangle \otimes |V\rangle \otimes |\omega\rangle)$$
$$\times (-i\langle\uparrow| \otimes \langle H| \otimes \langle\omega + \delta\omega| + \langle\downarrow| \otimes \langle V| \otimes \langle\omega|)$$

7.1 Spin-Photon Entanglement: Λ-System Decay

$$= \frac{1}{2}\Big[|\uparrow\rangle\langle\uparrow|\otimes|H\rangle\langle H|\otimes|\omega+\delta\omega\rangle\langle\omega+\delta\omega| + |\downarrow\rangle\langle\downarrow|\otimes|V\rangle\langle V|\otimes|\omega\rangle\langle\omega|$$
$$+ i|\uparrow\rangle\langle\downarrow|\otimes|H\rangle\langle V|\otimes|\omega+\delta\omega\rangle\langle\omega| - i|\downarrow\rangle\langle\uparrow||V\rangle\langle H|\otimes|\omega\rangle\langle\omega+\delta\omega|\Big]$$

(7.2)

The presence of coherences in the density matrix (off-diagonal elements) indicates entanglement, as opposed to a classical, mixed state:

$$\rho_{\text{mixed}} = \frac{1}{2}\Big[|\uparrow\rangle\langle\uparrow|\otimes|H\rangle\langle H|\otimes|\omega+\delta\omega\rangle\langle\omega+\delta\omega| + |\downarrow\rangle\langle\downarrow|\otimes|V\rangle\langle V|\otimes|\omega\rangle\langle\omega|\Big]$$

(7.3)

However, when considering only the polarization part of the photonic wavefunction, e.g. as a polarization-only qubit, tracing over the frequency-parts of the entangled state yields the following density matrix:

$$\rho_{\text{entangled,pol.}} = \text{Tr}_{\text{freq.}}(\rho_{\text{entangled}}) = \tfrac{1}{2}(|\uparrow\rangle\langle\uparrow|\otimes|H\rangle\langle H| + |\downarrow\rangle\langle\downarrow|\otimes|V\rangle\langle V|).$$

(7.4)

This is a mixed state in polarization and spin, *not* an entangled state. If one lets the environment trace over the frequency degrees of freedom, the leaked which-path information can destroy the entanglement. This is exactly the situation that occurs when considering spin-polarization entanglement only, using a measurement scheme that 'ignores' the frequency information. If that measurement scheme is, in principle, capable of detecting the difference, $\delta\omega$ between the two frequencies, which-path information will be leaked to the environment, and the entanglement measurement will not be able to distinguish it from a classical, mixed state. For a difference, $\delta\omega$, between the frequency components, the capability to distinguish this difference imposes a detector resolution (bandwidth) of better than $\delta\omega$, or, by virtue of the Fourier transformation, in the time-domain, a timing resolution of less than $1/\delta\omega$. Conversely, for broadband (fast, compared to $1/\delta\omega$) detectors, it is impossible to distinguish between the different frequency components. This is the basis of fast-detection based quantum-eraser schemes [3, 17, 18].

7.1.2 Quantum Erasure by Time-Resolved Measurement

The effect of the time-resolution of the measurement process on the spin-photon-polarization entanglement visibility can be clarified by considering what happens after detection of a photon of particular polarization, say, (σ^+) at time t_0. We obtain the following state for the spin:

$$|\Psi(t)_{\text{spin},t_0}\rangle = \frac{1}{\sqrt{2}}(e^{i\delta\omega(t-t_0)}|\uparrow\rangle - |\downarrow\rangle)$$

(7.5)

Moving to a density matrix description, the spin state corresponds to the following spin-density matrix:

$$\rho(t)_{\text{spin},t_0} = \frac{1}{2}\Big[|\uparrow\rangle\langle\uparrow| + |\downarrow\rangle\langle\downarrow|$$
$$+ ie^{i\delta\omega(t-t_0)}|\uparrow\rangle\langle\downarrow| - ie^{-i\delta\omega(t-t_0)}|\downarrow\rangle\langle\uparrow|\Big]. \quad (7.6)$$

Now, uncertainty over the measurement time t_0 gives rise to an ensemble average over different possible measurement times, which we can characterize by a distribution function $F(t_0)$:

$$\rho(t)_{\text{spin}} = \int \rho(t)_{\text{spin},t_0} F(t_0) dt_0. \quad (7.7)$$

This uncertainty is limited by both the accuracy (timing jitter) of the detector, and the spontaneous emission lifetime of the excited state of the Λ-system. For a very slow detection process, compared to the Larmor period, $1/\delta\omega$, we have that

$$\int e^{-i\delta\omega(t_0)} F(t_0) dt_0 \simeq \int_{-\infty}^{\infty} e^{-i\delta\omega(t_0)} dt_0$$
$$= 0 \quad (7.8)$$

such that

$$\rho(t)_{\text{spin,slow}} = \frac{1}{2}[|\uparrow\rangle\langle\uparrow| + |\downarrow\rangle\langle\downarrow|]. \quad (7.9)$$

The latter is a purely mixed state, not a pure state: slow measurement will select an ensemble of possible photon arrival times, each corresponding to spin states which have a totally different position in the equator of the Bloch sphere due to Larmor precession. Ensemble averaging over the arrival time of the photon will then destroy any coherence. Alternatively, in a quantum jump description [19], inaccuracy over the arrival time of the photon at the detector reflects itself in inaccuracy over the when exactly the quantum jump from the trion to the spin-ground states happened (Fig. 7.1), and therefore inaccuracy over the phase (time-origin) of the Larmor-precession of the spin. The ultimate uncertainty, for an infinitely slow detector, is then given by the spontaneous emission time of the excited trion state. Importantly, the same result would have been obtained had we started from a mixed spin-photon state rather than an entangled state. Slow detection can therefore not distinguish between a mixed and an entangled spin-photon state.

Conversely, a very fast detection process can distinguish between mixed and entangled spin-photon states. Suppose the distribution function $F(t_0)$ is very much peaked compared to $1/\delta\omega$ (i.e. that the detection timing resolution is fast compared to the Larmor period), then the following is true:

$$\int e^{-i\delta\omega(t-t_0)} F(t_0) dt_0 \simeq \int_{-\infty}^{\infty} e^{-i\delta\omega(t-t_0)} \delta(t_0 - \tilde{t_0}) dt_0$$
$$= e^{i\delta\omega(t-\tilde{t_0})}. \quad (7.10)$$

This results in the following density matrix for the spin state:

$$\rho(t)_{\text{spin,fast}} = \frac{1}{2}\Big[|\uparrow\rangle\langle\uparrow| + |\downarrow\rangle\langle\downarrow| \\ + ie^{i\delta\omega(t-\tilde{t}_0)}|\uparrow\rangle\langle\downarrow| - ie^{-i\delta\omega(t-\tilde{t}_0)}|\downarrow\rangle\langle\uparrow|\Big]. \quad (7.11)$$

The latter is now a pure state, not a mixed state, and different from what would be obtained for the case of a mixed spin-photon state. In the quantum jump picture, fast detection allows for selecting of only those quantum jump events that correspond to a very accurately defined time-slot, $\tilde{t}_0$. Therefore, the phase (time-origin) of the Larmor precession can be determined accurately. This, of course, corresponds to a form of time-filtering: of all the possible quantum jumps from the excited trion state, characterized by the spontaneous-emission decay in a time-ensemble measurement, we select only a very narrow time-slice – we refer to Appendix F for more details.

7.2 Ultrafast Frequency Downconversion as Quantum Eraser

For the ultrafast control techniques described in Chaps. 3 and 6, Larmor precession frequencies of several 10s of GHz were needed in order to measure the spin state accurately. Quantum erasing such difference in frequency would require a timing resolution of 10 ps or less, which no commercial single photon detector can provide. Our solution therefore consists of using a time-resolved, non-linear frequency downconversion technique, that mixes a spontaneously emitted photon with a short, broadband pulse in a difference-frequency generation (DFG) process [20]. In combination with narrowband filtering of the downconverted photons (to eliminate spurious light at any of the input frequencies), this provides an effective, accurate measurement of the arrival time of the spontaneously emitted photon at the downconverter: only those photons that arrive at exactly the same time as the short pulse can be converted.

Figure 7.2 displays the conversion setup used. For the quantum dot under study, the spontaneous emission decay is at 910 nm, and the target wavelength is 1,560 nm, corresponding to the lowest-loss band for telecom-fibers [12]. The required mixing frequency for such a process is at 2.2 µm, and short 2.2-µm light pulses can be generated by another DFG process that mixes 3-ps, 911-nm pulses from a modelocked laser with narrowband, continuous-wave (CW) 1,560-nm light in an MgO-doped, periodically poled lithium niobate (PPLN) chip (the efficiency of this process is improved significantly by modulating the 1,560 nm light by a fiber-optic modulator, and subsequently amplifying it by Erbium-doped fiber amplifiers). After wave mixing, the residual 1,560- and 911-nm light can be filtered out through a combination of dichroic and absorptive filters.

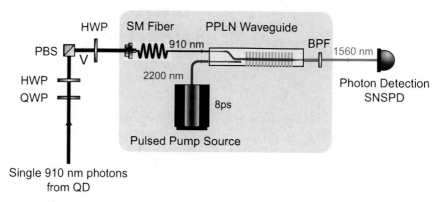

Fig. 7.2 Ultrafast downconversion for quantum erasure. See text for details. *QWP* quarter-waveplate, *HWP* half-waveplate, *SM fiber* single-mode fiber, *PPLN* periodically poled lithium-niobate, *SNSPD* superconducting nanowire single-photon detector. Figure reproduced from [16]

The downconversion process itself takes place in a PPLN waveguide [21] that confines the optical fields to a small region, and significantly improves the conversion-efficiency over bulk-systems: this is crucial for single-photon level conversion. Both the 910 nm photon and the 2.2 μ pulse are fiber-coupled to the inputs of the waveguide, and the residual scattered light from those inputs after conversion is eliminated through a combination of a fiber-Bragg grating and a long-pass filter, which form an effective bandpassfilter for the 1,560 nm single photon resulting from the downconversion process. This 1,560 nm photon can subsequently be detected on a superconducting nanowire single-photon detector (SNSPD); because of the mixing process, detecting a single 1,560 nm photon corresponds to detecting the arrival of a single 910 nm photon that exactly overlaps in time with the 2.2 μm pulse. In view of the polarization-selective operation of the downconversion process, the polarization measurement occurs by means of waveplates and polarizers *before* the downconversion process (see Fig. 7.2): only those photons with V-polarization right before the polarizer will be detected by the downconversion setup. The quarter- and halfwaveplates allow for different mappings of the emitted polarization into the measured V-polarization, and therefore determine the polarization-measurement basis used.

Figure 7.3 illustrates the performance of the downconversion technique. In Fig. 7.3a, we infer the timing resolution by mixing the 2.2 μm light with a short (3 ps), classical light pulse at 910nm. The resulting cross-correlation suggests a timing resolution of about 8 ps, limited by the duration of the 2.2μm pulse. This is in line with theoretical calculations, and the results are somewhat sensitive to the exact parameters used – in particular, the 2.2 μm pulse width depends quite strongly on the exact power and wavelength used for the DFG process, but is consistently between 3 and 8ps.

7.2 Ultrafast Frequency Downconversion as Quantum Eraser

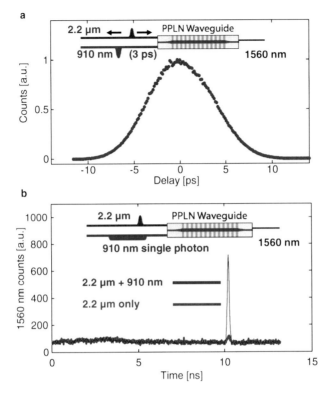

Fig. 7.3 Performance of the ultrafast downconversion technique. (**a**) Timing resolution, measured by cross-correlating the 2.2-µm conversion pulse with a classical, 3-ps pulse at 910 nm. From these data, we can infer a sub-8-ps resolution for the arrival time of a single-photon from the quantum dot. (**b**) Performance at single-photon levels, measured using a superconducting nanowire single photon detector (SNSPD). For single, 910-nm photons at the input (*blue trace*), the residual noise (*red trace*) can be seen to be well below the single-photon level (Figure reproduced from [16])

Figure 7.3b confirms the single-photon operation of the downconversion setup: in the absence of single, 910 nm photons at the input, the noise level is about 4–10 times lower than the signal level with 910 nm photons at the input. This noise floor is dominated by residual 2.2 µm leakage: in view of the 100 ps timing resolution of the SNSPD, the dark counts (some $40\,\mathrm{s}^{-1}$) can be restricted to this 100 ps timing window only. Such time-gating of the detectors can be performed in post-processing (in our case, based on analysis of the signal from a timing analyzer (Hydraharp, PicoQuant GmbH), used in time-tagged time-resolved (TTTR) mode), which permits dark-count suppression to well below the level of the residual 2.2 µm leakage.

7.3 Ultrafast Frequency Downconversion: Measurement of 910 nm Spin-Photon Entanglement

We use a single-electron-charged quantum dot for the verification of spin-photon entanglement (the single-hole case should be equivalent, in view of the similar selection rules, see Chap. 6). We refer to Fig. 7.4. For a magnetic field of 3 T, the Larmor precession of 17.6 GHz corresponds to a precession period of 57 ps. With a fixed timing resolution determined by the downconversion setup of 8 ps, this yields a theoretical entanglement visibility of about 95 % (fringe visibility of 90 % in the rotated basis – see Appendix F for details).

In order to verify entanglement, both the spin and the photon need to be measured in different bases. The photon polarization measurement occurs through waveplates and a polarizer in front of the downconversion setup, as reported below. The spin can be measured in different bases by a combination of a spin rotation and a measurement. The spin rotation technique is based on the ultrafast rotations presented in Chap. 3; by changing the spin rotation angle between 0, $\pi/2$ and π, followed by a measurement of the $|\downarrow\rangle$-state with an optical-pumping laser along the $|\downarrow\rangle - |\uparrow\downarrow\Downarrow\rangle$-transition (see Fig. 7.4a), we can distinguish between the $|\downarrow\rangle, |\uparrow\rangle$ and $|\rightarrow\rangle$-states (the measurement of the $|\leftarrow\rangle$-state involves delaying the $\pi/2$-pulse by half a Larmor period: this rotates the spin to the $|\rightarrow\rangle$-state, after which the $\pi/2$-pulse maps it into the $|\downarrow\rangle$-state). The fidelity of the initialization process and the spin rotations can be measured to be around 95 % or higher – we refer to Appendix A.1 for further details.

The optical setup used combines the ultrafast spin-control setup of Fig. 3.7 with the ultrafast downconversion setup of Fig. 7.2, and is indicated in Fig. 7.5. After initialization into the $|\downarrow\rangle$-state by optical pumping followed by a π-spin-rotation,

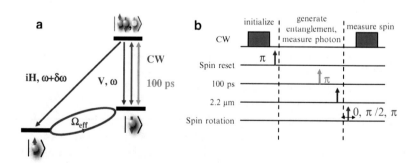

Fig. 7.4 Schematic of the procedure used to verify spin-photon entanglement. (**a**) Energy level and optical/spin transitions used. After initialization into the $|\downarrow\rangle$-state by a combination of optical pumping and a spin-π-pulse, a 100-ps optical π-pulse excites the system into the $|\uparrow\downarrow\Downarrow\rangle$-state. The polarization of the subsequently emitted single photon is measured by ultrafast downconversion, after which the spin is measured by a combination of an ultrafast spin-rotation pulse, and an optical-pumping pulse. (**b**) Timing of the respective steps in the procedure (Figures reproduced from [16])

7.3 Ultrafast Frequency Downconversion...

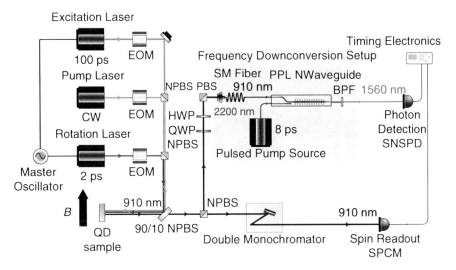

Fig. 7.5 Spin photon entanglement verification: full system diagram. See text for details. *EOM* electro-optic-modulator, *(N)PBS* non-polarizing beamsplitter (Figure reproduced from [16])

we apply a 100 ps optical π-pulse which excites the system into the $|\uparrow\downarrow\Downarrow\rangle$-trion state (excitation into the other trion state is prevented by accurate polarization control and use of the optical selection rules of the double Λ-system). This optical π-pulse is derived by pulse picking (we use a combination of free-space and fiber-based electro-optical modulators (EOM)) from a modelocked laser at 910 nm, resonant with the $|\downarrow\rangle$-$|\uparrow\downarrow\Downarrow\rangle$-transition. This modelocked laser is synchronized to the 911 nm modelocked laser that is used for both spin rotations (detuning: about 1 nm) and as seed laser for the generation of the 2.2 µm pulses used for time-resolved downconversion.

The subsequent spontaneous emission is collected by the confocal-microscopy setup, and split off on an NPBS. One branch (50 % probability of success) goes to the polarization-resolved downconversion setup, and is used for time-resolved measurement of the polarization of the photon: this is the second step in the schematic timing diagram of Fig. 7.4b.

In the next step, the spin-state is measured by the aforementioned combination of spin rotations and optical pumping; the emitted photon during the spin-measurement cycle is split off on the NPBS (50 % probability of success), and after cross-polarization and frequency-filtering measured on a 910 nm single-photon detector (SPCM). A timing analyzer in time-tagged time-resolved mode (Hydraharp, Pico-Quant GmbH, few-ps timing resolution) records all of the detected photons, both those measured by the SNSPD for measurement of the photon polarization, as those on the SPCM that quantify the spin-state. The raw datastream from the timing-analyzer can be transformed into spin-photon histograms, from which we can derive the probability of measuring the spin in a particular state, conditioned on a particular polarization of the photon.

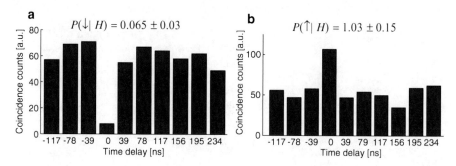

Fig. 7.6 Spin-photon correlation histograms, obtained after post-processing of the coincidence counts of spin photon on the timing analyzer. See text for details. (**a**) Correlation histogram for a measurement of the $|\downarrow\rangle$-spin-state and an H-photon. (**b**) Idem, for $|\uparrow\rangle$ and H (Figures reproduced from [16])

7.3.1 Histogram Analysis

The histograms, such as the one shown in Fig. 7.6, compare the coincidences between a downconverted single photon and a subsequent spin-measurement single photon, with coincidences between downconverted photons and spin-measurements occurring in different cycles of the experiment. The latter act as an effective normalization procedure, as we will show further. The analysis relies on three assumptions, which can all be verified: (1) the probability of photon emission along the different branches of the Λ-systems is identical; (2) there is no correlation between events occurring in different cycles of the experiment; and (3) the probability of a photon being detected does not vary appreciably within a few cycles of the experiment (no fast blinking of the quantum dot).

The first assumption relies on the oscillator strengths of all Λ-system transitions being equal, and can be easily verified using e.g. magnetophotoluminescence. As the intensities of all four transitions in the double Λ-system are equal, their oscillator strengths do indeed not vary appreciably.

The second assumption is correct, provided that there is proper re-initialization of the spin after each cycle, which is realized by a combination of optical pumping into the $|\uparrow\rangle$-state, followed by a spin-rotating π-pulse to bring the system into the $|\downarrow\rangle$-state. With initialization and rotation pulse fidelities $\geq 95\%$ as derived above, any hidden correlation between events in subsequent experimental cycles should be well suppressed below 10%.

The third assumption is more difficult to calibrate, and can also affect the normalization procedure. If, due to blinking or other memory effects in or nearby the quantum dot, the charge state or spectral position were to change from one cycle of the experiment to the other, then the photon extraction probability would vary from cycle to cycle. This would make comparison of coincidences between events within the same cycle of the experiment and subsequent cycles problematic. While (anti)correlations as fast as 10 ns have been observed previously in InAs quantum

7.3 Ultrafast Frequency Downconversion...

dot photoluminescence using above-band or quasi-resonant excitation [22], we do not observe any of the characteristic, systematic histogram variances in our experiments, where we use on-resonance excitation. While it is possible that small memory effects do exist, they are dominated by the Poissonian variance of our data, and are therefore within the error bars of our analysis.

Each experimental run takes either 39 or 52 ns, and consists of the three cycles described above (Fig. 7.4b): (1) initialization of the spin; (2) optical excitation and generation of the spin-photon entanglement, followed by detection of the downconverted single photon at 1,560 nm with the superconducting detector (SNSPD); (3) spin rotation and measurement, through detection on the single photon counter (SPCM) of the single 910 nm photon emitted in the optical pumping experiment. The last part of (3) overlaps with the optical pumping part of (1), and the entire run is then repeated for several hours.

After several hours of integration (approximately 500 MB of data per hour of integration), we load the obtained datastream into a custom C++-program. Given the accurate timing of the experiment, it is possible to eliminate a large fraction of (spurious) data due to detector dark counts: it is only possible for the physical detection signals of both spin and photon to arrive at well defined times. We therefore start our analysis by excluding all events outside certain well-defined time windows, which comes down to a very effective form of gating in post-processing. In particular, the downconverted single photons have a duration of a few ps, and the SNSPD detector a timing response of about 100 ps. Only considering detection events within these 100 ps reduces the amount of dark counts from the SNSPD by a factor of 400–500 (100 ps vs. 39 or 52 ns for the duration of the entire experiment). Likewise, the SPCM signals are also gated, to reduce dark counts during the parts of the experiment when the spin signal is not monitored (13 ns gating time vs. 39 or 52 ns). Other than the dark-count removal outside the time-window of interest, no other corrections are applied to the data (no background subtraction of any kind). This is notably different form the experiments in Chaps. 3 and 6: those were performed using a digital lock-in procedure. Improved cross-polarization and filtering allow us to remove this lock-in procedure and obtain high signal-to-noise measurements of the spin state.

From this filtered data-stream, we then calculate a correlation histogram by conditioning on the arrival time of a downconverted photon. We look for correlations within the same experimental run, and compare those to events occurring in neighboring experimental runs (delaying the spin signal electronically allowed for inclusion of correlations at negative time delays). With equal decay probabilities within the Λ-system, and perfect reinitialization of the spin in a subsequent cycle, the expected coincidence rate for events in subsequent cycles is 50%, which allows for normalization of the coincidence rate within the same cycle. With the losses incurred in the experiment (0.1% single photon detection efficiency; downconversion losses; time-filtering of a sub-10 ps window from spontaneous emission decay with 600 ps decay time), we obtain a downconverted single-photon detection rate of about 2–5 Hz for a 52 ns cycle time, and an average coincidence rate of 2–5 mHz due to the 0.1% probability of detecting the subsequent, spin-measuring

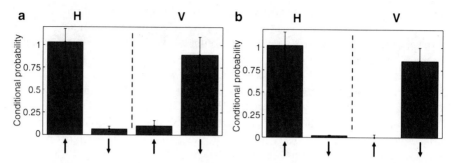

Fig. 7.7 Spin-photon correlations in the linear (eigen) basis. (**a**) Conditional probability of measuring the spin in a particular state, conditional on measurement of the photon in the linear (H,V) basis. (**b**) Same, without downconversion to 1,560 nm (Figures adopted from [16])

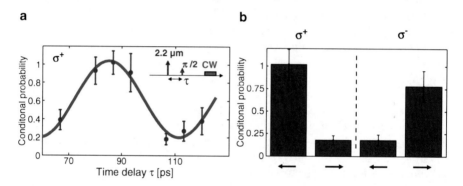

Fig. 7.8 Spin-photon correlations in the rotated basis. (**a**) Upon detection of a σ^+-photon, the spin starts precessing due to Larmor precession. Application of a $\pi/2$ pulse at different times permits tracing out of this coherent precession, and measurement of the (anti-)correlations. (**b**) Conditional probability of measuring the spin in a particular state, conditional on measurement of the photon in the circular basis ($\sigma^\pm$) (Figures reproduced from [16])

single photon. With several hours of integration, the total number of coincidences reaches up to about 50–100, with the variances on the coincidence rates consistent with Poissonian statistics.

The main systematic source of error in our experiment, which is not subtracted from the data, is due to uncorrelated light/detection events at 1,560 nm. These consist of dark counts within the 100 ps response time window of the SNSPD, residual leakage from the 2.2 µm pump pulse, and residual leakage from the 100 ps excitation pulse. For the data in Figs. 7.6 and 7.7 (computational/non-rotated basis), the residual 100 ps leakage could be reduced due to cross-polarization, leading to a signal-to-noise ratio of about 10:1. For the data in Fig. 7.8 (rotated basis of spin and photon), only time-filtering could be used to remove the effect of the 100 ps laser reflections (the spontaneous emission has a decay time of approximately 600 ps, allowing a suppression by about 40 dB of the reflected pump laser light by moving the time-window for downconversion by approximately 500 ps after the excitation pulse), leading to a signal-to-noise ratio of about 4:1.

7.3.2 Linear Basis: Correlations

For the linear (computational/non-rotated basis), the correlations are shown in Fig. 7.7a: for an H-polarized photons measured after downconversion, the correlation with the $|\uparrow\rangle$-state is very good. Likewise, detection of a V-polarized photon results in the spin being found in the $|\downarrow\rangle$-state. These results, obtained with ultrafast downconversion, can be compared with a direct measurement of the 910 nm photon using an SPCM: Fig. 7.7b. For both cases, the correlations are in excellent agreement; the small differences are explained by the residual noise present in the downconversion process.

7.3.3 Rotated Basis: Entanglement

Verification of entanglement requires observation of correlations in a rotated basis of the photon polarization and the spin as well. When we measure the photon in the circular polarization basis ($|\sigma^+\rangle$, $|\sigma^-\rangle$), we measure the spin in the basis of $|\rightarrow\rangle = \frac{1}{\sqrt{2}}(|\uparrow\rangle + |\downarrow\rangle)$ and $|\leftarrow\rangle = \frac{1}{\sqrt{2}}(|\uparrow\rangle - |\downarrow\rangle)$. After detection of a $|\sigma^+\rangle$ ($|\sigma^-\rangle$)-downconverted photon at time t_1, the spin is projected into the $|\leftarrow\rangle$-state ($|\rightarrow\rangle$), which subsequently evolves in time due to Larmor precession:

$$|\Psi_{\text{spin}}(t)\rangle = \frac{1}{\sqrt{2}}(e^{i(\delta\omega)(t-t_1)}|\uparrow\rangle \mp |\downarrow\rangle). \tag{7.12}$$

Here $\delta\omega$ corresponds to the Zeeman frequency of $2\pi \times 17.6\,\text{GHz}$. By scanning the arrival time of a $\pi/2$ spin rotation pulse in a Ramsey interferometer [23, 24] (see Fig. 7.8a) we can trace out this coherent oscillation, and verify entanglement. From the minima and maxima of these coherent oscillations, we can derive, as in Fig. 7.8b, the photon-spin correlations for $|\sigma^+\rangle$ ($|\sigma^-\rangle$) downconversion. For a particular photon polarization ($|\sigma^+\rangle$, $|\sigma^-\rangle$) and arrival time of the $\pi/2$ spin rotation pulse, a correlation can be measured, as expected. Subsequently changing the arrival time of the $\pi/2$-pulse by half a Larmor period results in an anticorrelation.

By combining the histograms for different experimental configurations, we can extract the experimental fidelity. The entanglement fidelity analysis follows the same procedure used in the ion-trap [4] and NV-diamond [3] spin-photon entanglement experiments:

$$F \geq F_1 + F_2,$$

$$F_1 = \frac{1}{2}(\rho_{H\uparrow,H\uparrow} + \rho_{V\downarrow,V\downarrow} - 2\sqrt{\rho_{H\downarrow,H\downarrow}\rho_{V\uparrow,V\uparrow}}),$$

$$F_2 = \frac{1}{2}(\rho_{\sigma^+\leftarrow,\sigma^+\leftarrow} - \rho_{\sigma^+\rightarrow,\sigma^+\rightarrow} + \rho_{\sigma^-\rightarrow,\sigma^-\rightarrow} - \rho_{\sigma^-\leftarrow,\sigma^-\leftarrow}).$$

Here $\rho_{H\uparrow,H\uparrow}$ etc. refer to elements of the spin photon density matrix, which can be associated with the observed correlations in our experiment. As the emission rate of H- and V-polarized photons is equal, as we argued before, we can associate the following matrix elements with the obtained conditional probabilities [3]:

$$\rho_{H\uparrow,H\uparrow} = \frac{1}{2}P(\uparrow|H) = \frac{1}{2}(1.03 \pm 0.15)$$

$$\rho_{V\downarrow,V\downarrow} = \frac{1}{2}P(\downarrow|V) = \frac{1}{2}(0.89 \pm 0.2)$$

$$\rho_{H\downarrow,H\downarrow} = \frac{1}{2}P(\downarrow|H) = \frac{1}{2}(0.065 \pm 0.03)$$

$$\rho_{V\uparrow,V\uparrow} = \frac{1}{2}P(\uparrow|V) = \frac{1}{2}(0.1 \pm 0.06)$$

$$\rho_{\sigma^+\leftarrow,\sigma^+\leftarrow} = \frac{1}{2}P(\leftarrow|\sigma^+) = \frac{1}{2}(1.02 \pm 0.17)$$

$$\rho_{\sigma^-\rightarrow,\sigma^-\rightarrow} = \frac{1}{2}P(\rightarrow|\sigma^-) = \frac{1}{2}(0.78 \pm 0.17)$$

$$\rho_{\sigma^+\rightarrow,\sigma^+\rightarrow} = \frac{1}{2}P(\rightarrow|\sigma^+) = \frac{1}{2}(0.18 \pm 0.05)$$

$$\rho_{\sigma^-\leftarrow,\sigma^-\leftarrow} = \frac{1}{2}P(\leftarrow|\sigma^-) = \frac{1}{2}(0.18 \pm 0.07)$$

From these numbers, we obtain that $F_1 = 0.44 \pm 0.06$ and $F_2 = 0.36 \pm 0.06$, resulting in an experimental fidelity of 0.80 ± 0.085, which is limited by the residual leakage in the downconversion channel, and does not reach the theoretical upper limit of 0.95 (F_2 limited to 0.907/2 due to the finite timing resolution – see Appendix F). This experimental fidelity does, however, exceed the classical threshold of 0.5 by more than 3 standard deviations, and unambiguously demonstrates spin-photon entanglement in our system.

7.4 Towards 1,560 nm Spin-Photon Entanglement by Ultrafast Downconversion

By using the ultrafast downconversion as a time-resolved measurement tool for the polarization of the 910 nm photon, we can verify spin-photon entanglement. However, by adding a second downconversion setup and simple, linear elements, a full photonic qubit at 1,560 nm could be realized, and entanglement could be extended to the 1,560 nm telecom band (resource limitations, in particular the availability of sufficient 2.2 μm light power, prevented us from running two downconversion setups in parallel at the time of writing of this dissertation, but this limitation is practical, not fundamental).

7.4 Towards 1,560 nm Spin-Photon Entanglement by Ultrafast Downconversion

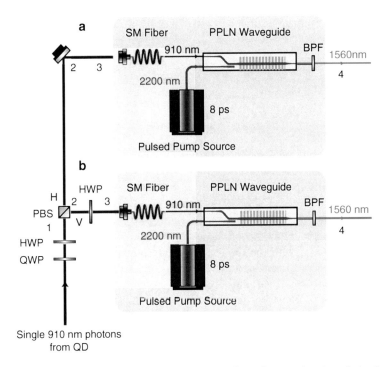

Fig. 7.9 Transforming spin-photon entanglement to 1,560 nm. By mapping the polarization into a dual-rail degree of freedom, we can quantum erase the frequency which path information and obtain a 1,560 nm photonic qubit, entangled with the spin. See text for details (Figure adopted from [16])

The main complication for realizing a full photonic qubit at 1,560 nm, is the strong polarization-dependence of the PPLN waveguides used. Effective downconversion requires the polarization to be aligned along a particular direction with regard to the plane of the waveguide (we refer to this polarization as H). By including a second rail in the downconversion setup, the polarizing beamsplitter can map the different polarization states to different branches of the dual rail. We start from position 1 in Fig. 7.9. The quantum state, including the color of the photons, can now be described as follows:

$$|\Psi_1\rangle = \frac{1}{\sqrt{2}}(i|\uparrow\rangle \otimes |H\rangle \otimes |\omega + \delta\omega\rangle + |\downarrow\rangle \otimes |V\rangle \otimes |\omega\rangle). \tag{7.13}$$

The polarizing beamsplitter now sends H-polarized photons to branch a, and V-polarized ones to branch b (position 2):

$$|\Psi_2\rangle = \frac{1}{\sqrt{2}}(i|\uparrow\rangle \otimes |H\rangle \otimes |a\rangle \otimes |\omega + \delta\omega\rangle + |\downarrow\rangle \otimes |V\rangle \otimes |b\rangle \otimes |\omega\rangle). \tag{7.14}$$

With the half-waveplate in branch b (in order to match the polarization of the waveguide), we obtain the following state for position 3:

$$|\Psi_2\rangle = \frac{1}{\sqrt{2}}(i|\uparrow\rangle \otimes |H\rangle \otimes |a\rangle \otimes |\omega + \delta\omega\rangle + |\downarrow\rangle \otimes |H\rangle \otimes |b\rangle \otimes |\omega\rangle). \qquad (7.15)$$

Here, both branches are H-polarized.

Now, for ultrafast conversion at time t_0 (we assume that both branches are synchronized; if not, then an additional, branch-dependent phase factor develops, which does not affect the physics other than that it creates a different entangled state), the generated photons at ω_{1560} have a bandwidth of about $1/8$ ps ~ 100 GHz, which significantly exceeds the initial difference in wavelength $\delta\omega$ between the respective photonic partial waves. These photons are therefore indistinguishable: as argued above, this is the essence of the quantum eraser technique [17]. Note that this interpretation of the ultrafast conversion in the frequency domain is perfectly compatible with the time-filtering interpretation given above: both are related by Fourier transformations. In either case, the entanglement obtained is post-selective, and valid for a particular subset of the spontaneously emitted photons. Because of the quantum indistinguishability, we obtain the following state (position 4 in Fig. 7.9):

$$|\Psi_4(t)\rangle = \frac{1}{\sqrt{2}}(ie^{i\delta\omega(t-t_0)}|\uparrow\rangle \otimes |H\rangle \otimes |a\rangle \otimes |\omega_{1560}\rangle + |\downarrow\rangle \otimes |H\rangle \otimes |b\rangle \otimes |\omega_{1560}\rangle).$$

$$(7.16)$$

We can now omit the polarization and wavelength labels of the photonic partial waves (they are common to both branches), and obtain a new state, explicitly entangled only in the path or rail (a, b) degree of freedom. The photonic part of this entangled state exists at telecom wavelengths (1,560 nm):

$$|\Psi_4(t)\rangle = \frac{1}{\sqrt{2}}(ie^{i\delta\omega(t-t_0)}|\uparrow\rangle \otimes |a\rangle + |\downarrow\rangle \otimes |b\rangle) \qquad (7.17)$$

By adding another half-waveplate and a polarizing beamsplitter, we could recombine both rail-based partial photonic waves into a 1,560 nm, polarization qubit (see Fig. 7.10):

$$|\Psi_5(t)\rangle = \frac{1}{\sqrt{2}}(ie^{i\delta\omega(t-t_0)}|\uparrow\rangle \otimes |H\rangle + |\downarrow\rangle \otimes |V\rangle) \qquad (7.18)$$

Alternatively, it is possible to delay one branch with regard to the other, and recombine both branches on a beamsplitter or switch without changing either polarization: this way, time-bin entanglement could be created, again at telecom wavelengths [25] – the latter would be preferable over polarization entanglement in view of the potential distortion of the polarization in commercial telecommunication fibers. Yet another possibility, further removed from the present work, would be

7.4 Towards 1,560 nm Spin-Photon Entanglement by Ultrafast Downconversion

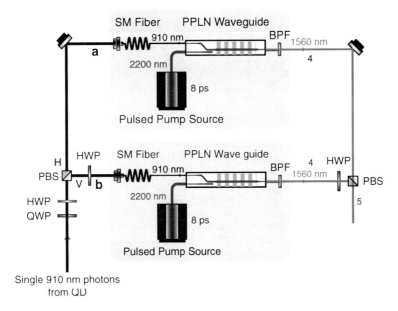

Fig. 7.10 Realization of a 1,560 nm, polarization entangled qubit by recombining the dual-rail qubit in Fig. 7.9. See text for details (Figure reproduced from [16])

to replace the ultrafast, pulsed downconverters presented here with continuous-wave (CW) versions, pumped by the same CW-laser source at 2.2 μm. In such a scheme, quantum erasure of the frequency difference would not occur, and the downconverted, 1,560 nm photons would have a frequency difference, $\delta\omega$, that is equal to the difference before conversion. By recombining the partial waves (which would then both have the same polarization, as in the scheme described above, such that polarization which-path information is effectively erased), one could obtain frequency-entanglement at 1,560 nm, which is more robust against fiber-fluctuations. An added benefit of such a scheme would be in the significantly higher photon yield, as no time-filtering occurs. Verification of such frequency entanglement, however, would be challenging for the large Zeeman energy splittings used in our system.

By combining two spin-photon entangled pairs, and interfering them on a beamsplitter in a Hong-Ou-Mandel-configuration [26], one could realize entanglement swapping and projective, probabilistic remote spin entanglement, as we argued in Sect. 1.1.3.2. Moreover, having the photonic part of the entangled spin-photon pair at the lowest-loss wavelength would allow the intermediate nodes in a future quantum repeater to be spaced relatively far apart [12] – we refer to Sect. 1.4 for more details. The Hong-Ou-Mandel scheme relies on the photonic partial waves of the respective spin-photon entangled pairs being indistinguishable. Such indistinguishability has been observed for quantum dot photons around 900 nm [18], but the visibility is limited due to fluctuations in the solid-state environment.

As was shown in that work and in others [27], fast detection could improve the indistinguishability, and therefore increase the fidelity of potential entanglement swapping schemes. Our time-resolved conversion technique could perform a similar role, all the while transforming the spin-photon entanglement to low-loss fiber wavelengths.

References

1. H. J. Kimble. The quantum internet. *Nature*, 453:1023, 2008.
2. S. Ritter, C. Nölleke, C. Hahn, A. Reiserer, A. Neuzner, M. Uphoff, M. Mücke, E. Figueroa, J. Bochmann, and G. Rempe. An elementary quantum network of single atoms in optical cavities. *Nature*, 484:195, 2012.
3. E. Togan, Y. Chu, A.S. Trifonov, L. Jiang, J. Maze, L. Childress, M. V. G. Dutt, A. S. Sørensen, P. R. Hemmer, A. S. Zibrov, and M. D. Lukin. Quantum entanglement between an optical photon and a solid-state spin qubit. *Nature*, 466:730, 2010.
4. B. B. Blinov, D. L. Moehring, L.-M. Duan, and C. Monroe. Observation of entanglement between a single trapped atom and a single photon. *Nature*, 428:153, 2004.
5. T. Wilk, S. C. Webster, A. Kuhn, and G. Rempe. Single-atom single-photon quantum interface. *Science*, 317:488, 2007.
6. L.-M. Duan, M. D. Lukin, J. I. Cirac, and P. Zoller. Long-distance quantum communication with atomic ensembles and linear optics. *Nature*, 414:413, 2001.
7. H.-J. Briegel, W. Dür, J. I. Cirac, and P. Zoller. Quantum repeaters: The role of imperfect local operations in quantum communication. *Phys. Rev. Lett.*, 81:5932, 1998.
8. Z.-S. Yuan, Y.-A. Chen, B. Zhao, S. Chen, J. Schmiedmayer, and J.-W. Pan. Experimental demonstration of a BDCZ quantum repeater node. *Nature*, 454:1098, 2008.
9. A. Stute, B. Casabone, P. Schindler, T. Monz, P. O. Schmidt, B. Brandstätter, T. E. Northup, and R. Blatt. Tunable ion-photon entanglement in an optical cavity. *Nature*, 485:482, 2012.
10. D. L. Moehring et al. Entanglement of single-atom quantum bits at a distance. *Nature*, 449:68, 2007.
11. C. W. Chou et al. Measurement-induced entanglement for excitation stored in remote atomic ensembles. *Nature*, 438:828, 2005.
12. H. Takesue, S. W. Nam, Q. Zhang, R. H. Hadfield, T. Honjo, K. Tamaki, and Y. Yamamoto. Quantum key distribution over a 40-dB channel loss using superconducting single-photon detectors. *Nat. Photonics*, 1:343, 2007.
13. P. Michler, A. Kiraz, C. Becher, W. V. Schoenfeld, P. M. Petroff, L. Zhang, E. Hu, and A. Imamoglu. A quantum dot single-photon turnstile device. *Science*, 290:2282, 2000.
14. E. Moreau, I. Robert, L. Manin, V. Thierry-Mieg, J. M. Gérard, and I. Abram. A single-mode solid-state source of single photons based on isolated quantum dots in a micropillar. *Physica E*, 13:418, 2002.
15. M. Pelton, C. Santori, J. Vuckovic, B. Zhang, G. S. Solomon, J. Plant, and Y. Yamamoto. Efficient source of single photons: A single quantum dot in a micropost microcavity. *Phys. Rev. Lett.*, 89:233602, 2002.
16. K. De Greve, L. Yu, P. L. McMahon, J. S. Pelc, C. M. Natarajan, N. Y. Kim, E. Abe, S. Mier, C. Schneider, M. Kamp, S. Höfling, R. H. Hadfield, A. Forchel, M. M. Fejer, and Y. Yamamoto. Quantum-dot spin-photon entanglement via frequency downconversion to telecom wavelength. *Nature*, 491:421, 2012.
17. M. O. Scully and K. Drühl. Quantum eraser: A proposed photon correlation experiment concerning observation and "delayed choice" in quantum mechanics. *Phys. Rev. A*, 25:2208, 1982.

References

18. R. B. Patel, A. J. Bennett, I. Farrer, C. A. Nicoll, D. A. Ritchie, and A. J. Shields. Two-photon interference of the emission from electrically tunable remote quantum dots. *Nat. Photonics*, 4:632, 2010.
19. M. O. Scully and M. S. Zubairy. *Quantum optics*. Cambridge University Press, 1997.
20. R. W. Boyd. *Nonlinear optics*. Academic Press, 2003.
21. J. S. Pelc, L. Yu, K. De Greve, C. M. Natarajan, P. L. McMahon, *et al.* 1550-nm downconversion quantum interface for a single quantum dot spin. *in preparation*.
22. C. Santori, D. Fattal, J. Vuckovic, G. S. Solomon, E. Waks, and Y. Yamamoto. Submicrosecond correlations in photoluminescence from InAs quantum dots. *Phys. Rev. B*, 69:205324, 2004.
23. D. Press, T. D. Ladd, B. Zhang, and Y. Yamamoto. Complete quantum control of a single quantum dot spin using ultrafast optical pulses. *Nature*, 456:218, 2008.
24. D. Press, K. De Greve, P. McMahon, T. D. Ladd, B. Friess, C. Schneider, M. Kamp, S. Höfling, A. Forchel, and Y. Yamamoto. Ultrafast optical spin echo in a single quantum dot. *Nat. Photonics*, 4:367, 2010.
25. I. Marcikic, H. de Riedmatten, H. Zbinden, M. Legré, and N. Gisin. Distribution of Time-Bin Entangled Qubits over 50 km of Optical Fiber. *Phys. Rev. Lett.*, 93:180502, 2004.
26. C. K. Hong, Z. Y. Ou, and Mandel L. Measurement of Subpicosecond Time Intervals between Two Photons by Interference. *Phys. Rev. Lett.*, 59:2044, 1987.
27. C. Santori, D. Fattal, J. Vuckovic, G. S. Solomon, and Y. Yamamoto. Indistinguishible photons from a single-photon device. *Nature*, 419:594, 2002.

Chapter 8
Conclusion and Outlook

The work presented in this dissertation elaborates on a platform (optically controlled, charged, self-assembled InAs quantum dots) and a set of technologies for all-optical quantum information processing, with a specific emphasis on solid-state quantum repeaters. In Fig. 1.10, reproduced in Fig. 8.1, we indicated the elementary building blocks of a small, proof-of-principle quantum repeater.

8.1 High-Fidelity, Coherent Single Qubit Control

In Chap. 3, we reported on previous experiments [1,2] where the spin state of single, electron-charged quantum dots were manipulated by a combination of ultrafast coherent optical rotations, and magnetic-field induced Larmor precession. The main limitations on the fidelity of the control operations were found to be strong, nuclear hyperfine interactions [3], and rotation pulse errors due to the finite duration of the laser pulses used.

The rotation pulse errors were shown to affect the decoupling pulses used in spin-echo experiments, and a solution was presented in Chap. 4, where composite pulses based on Hadamard gates were used to significantly increase the fidelity of the decoupling pulses. We demonstrated a simple pulse-stretching technique, based on a double-grating pulse shaper [4], that can generate Hadamard gates with a single, optimized ultrafast pulse.

The limitations of the nuclear spin interactions on the control fidelity of electron spins were tackled in Chap. 6, where we demonstrated that spin-qubits based on hole spins do not suffer from the strong hyperfine interactions that electrons are subject too. While this does indeed increase the fidelity of Larmor-precession based control operations, unfortunately, hole spins were also shown to be more susceptible to unavoidable electric-field fluctuations [5]. While such electric field fluctuations

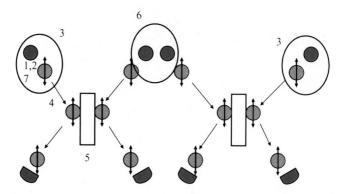

Fig. 8.1 Basic ingredients for a quantum repeater, as discussed in this work. *Green circles*: memory (spin) qubits; *orange circles*: single photonic qubits; *black-and-white rectangles*: beam-splitters for HOM-measurement; *green boxes*: single-photon detectors; *black-and-white circles*: entanglement operations

do not seem unavoidable (unlike the nuclear hyperfine interaction for electrons), further optimization would be required for hole spins to really outperform electrons, especially in terms of coherence times.

8.2 Long-Lived Quantum Memories

The coherence time of electron spin qubits was found to be on the order of several microseconds in Ref. [2], using a spin-echo technique. While we investigated hole spins as a possible alternative (see Chap. 6), their coherence was shown to be limited by electric-field fluctuations, with similar coherence times as the electron spin [5]. Further device engineering and optimization might lead to improved hole coherence times, as the hyperfine interactions were indeed shown to be significantly lower hole spins, in line with theoretical predictions [6].

Another approach to improve the spin coherence could consist of a higher-order spin-echo sequence, the so-called dynamical decoupling approach [7–9]. The improvements in π-pulse fidelity reported in Chap. 4 can be seen in this context: as the Hadamard-gate-based composite π-pulses significantly increase the effectiveness of a single-echo decoupling pulse, it is expected that multi-pulse decoupling schemes should benefit as well. Initial experiments do indeed show good visibility of the interference fringes in multi-pulse decoupling sequences. However, no improvements in observed coherence times have been observed yet in optically controlled quantum dots, in contrast to electrically controlled, gate-defined quantum dots [9] – this remains a subject under study.

Improving the coherence times of the spin qubits is crucial: with the current coherence times of several microseconds only, the longest possible distance over which spin-spin entanglement can be distributed is several hundred meters only! Without further improvement, quantum dot based quantum repeaters cannot be considered viable candidates. In principle, improvements up to the T_1 (spin-flip) time of several milliseconds could be possible (several 100 km), after which active error-correcting schemes [10] could provide architectural protection to the quantum state. Such approaches do, however, require significant advances in e.g. 2-qubit gate performance. Realistically, therefore, short-term future experiments will be inherently limited to short-distance distribution of spin-spin entanglement.

8.3 Spin-Photon Entanglement

In Chap. 7, we discussed a proof-of-principle experiment where entanglement between a single quantum dot electron spin and the polarization of a spontaneously emitted photon was demonstrated, with reasonably high fidelity (80 % and higher). The ultrafast frequency conversion technique used can be extended to generate spin-photon entanglement with a downconverted, telecom-wavelength photon (see Sect. 7.4). While the timing-resolution of the conversion-technique provides the quantum erasure necessary for observing spin-polarization entanglement, a variant using narrowband, CW downconversion and polarization erasure could result in more robust, frequency-entangled spin-photon pairs compatible with long-distance fiber communication.

Such spin-photon entanglement can now be used as a resource for future experiments involving photonic Bell-state measurements and entanglement swapping [11, 12]. In view of the relatively short spin coherence times, however, such entanglement swapping experiments will, at present, inevitably be limited to short distances.

8.4 Low-Loss Photonic Qubits

The ultrafast frequency-downconversion experiments reported in Chap. 7 can be extended to realize a full photonic qubit at telecommunication wavelengths (1,560 nm), either in a dual-rail implementation as in Fig. 7.9 or by recombining the dual-rail into a polarization or time-bin qubit at 1,560 nm (Fig. 7.10). In addition, the aforementioned CW-downconversion variant of our experiment would provide a fiber-compatible, frequency-entangled spin-photon pair.

The limitation of the spin coherence time does restrict the distance over which spin-photon entanglement can be distributed. However, the frequency-downconversion technique was demonstrated at the single-photon level, and could therefore be used for generating high-quality, fast, single photon sources at 1,560 m, which are by themselves useful for quantum-key-distribution [13, 14].

8.5 High-Fidelity Photonic Quantum Interference

The generation of remote-spin entanglement based on entanglement-swapping of spin-photon entangled pairs [12, 15] requires the implementation of a photonic Bell-state measurement. In Sect. 1.1.3.2, we illustrated a probabilistic version of such Bell-state measurement using linear optics and detector-post-selection [16]. The crucial feature used in such a scheme is the quantum interference between indistinguishable single photons [11]. Such indistinguishability was demonstrated for spontaneously emitted single photons in self-assembled quantum dots [17, 18], but those experiments were based on post-selecting quantum dots that emit photons at wavelengths very close to each other, and tuning them into resonance (typical quantum dots can emit in a wavelength range of over 100 nm). In addition, spectral wandering was shown to be limiting the visibility of the quantum interference.

While it remains to be demonstrated, quantum interference of downconverted single photons using the techniques demonstrated in Chap. 7 could tackle these issues. On the one hand, the conversion technique permits downconversion of different wavelengths to the same target wavelength by appropriate choice of the mixing wavelength, which would solve the inhomogeneous broadening of the quantum dots. Moreover, the temporal shaping (time-filtering) could offer a solution to the spectral wandering; in general, any fast detection/broadband technique can be shown to increase such quantum interference fidelity [18], at the expense of filtering and the resulting loss in count rates.

8.6 High-Fidelity, Entangling 2-Qubit Gate

A truly scalable, entangling 2-qubit gate for spins in self-assembled quantum dots remains one of the biggest outstanding challenges for this technology. It is crucial for both quantum error correction [10] that would be required for realistic repeaters, and for entanglement swapping between neighboring nodes [19]. A non-scalable, 2-qubit gate was demonstrated in Ref. [20] for quantum dot molecules, but this does not allow for arbitrary nearest-neighbor interactions to be established.

Several proposals have been published, based on differential geometric phases upon a common interaction with cavity fields coupling to neighboring quantum dots [21], but those require quite stringent parameter sets and have yet to be implemented. The work on single-qubit geometric phases presented in Chap. 5 does confirm that such geometric phases can be applied to quantum dot spins with relatively high-fidelity and on short timescales, and could therefore be seen as a primer to developing geometric-phase based 2-qubit gates.

8.7 High-Fidelity, Efficient Quantum Memory Readout

Another outstanding challenge is and remains the high-fidelity, fast, single-shot readout of a single qubit. While our current, single-photon based technique outlined in Sect. 3.2.1 does provide one bit of information (measurement of the $|\downarrow\rangle$-state) per detected photon, the probability of success of detecting such a photon is only on the order of 1 % or less – making the measurement technique effectively multi-shot.

Techniques based on the dispersive readout of a single photon in either Kerr- or Faraday rotation [22, 23] were demonstrated in multi-shot by accurate averaging, and may or may not be extended to single-shot implementations – this is the subject of ongoing research. An alternative approach used resonance-fluorescence from a quantum dot molecule [24], which was shown to yield single-shot readout, but only when working in the Faraday-geometry. The different selection rules in the Voigt geometry, which is in itself compatible with the optical Λ-systems required for fast optical spin control and spin-photon entanglement generation, seem incompatible with such a strategy.

8.8 Outlook

While significant advances have been made towards both all-optical control of quantum dot spins for quantum computing and quantum communication purposes, the realization of a small-scale demonstrator system, such as a small-scale quantum repeater or even a pair of remotely entangled quantum dot spins, will still require several years of dedicated research. At the time of writing of this dissertation, the crucial bottlenecks seem to be threefold: (1) limited coherence time of the qubits; (2) absence of a scalable 2-qubit gate; and (3) realization of a single-shot readout scheme, compatible with the Voigt geometry. Without a solution to all of them, self-assembled quantum dots will not be suitable as a system-technology for long-distance quantum key distribution.

References

1. D. Press, T. D. Ladd, B. Zhang, and Y. Yamamoto. Complete quantum control of a single quantum dot spin using ultrafast optical pulses. *Nature*, 456:218, 2008.
2. D. Press, K. De Greve, P. McMahon, T. D. Ladd, B. Friess, C. Schneider, M. Kamp, S. Höfling, A. Forchel, and Y. Yamamoto. Ultrafast optical spin echo in a single quantum dot. *Nat. Photonics*, 4:367, 2010.
3. T. D. Ladd, D. Press, K. De Greve, P. McMahon, B. Friess, C. Schneider, M. Kamp, S. Höfling, A. Forchel, and Y. Yamamoto. Pulsed nuclear pumping and spin diffusion in a single charged quantum dot. *Phys. Rev. Lett.*, 105:107401, 2010.
4. A. M. Weiner. Femtosecond pulse shaping using spatial light modulators. *Rev. Sci. Instrum.*, 71:1929, 2000.

5. K. De Greve, P. L. McMahon, D. Press, T. D. Ladd, D. Bisping, C. Schneider, M. Kamp, L. Worschech, S. Höfling, A. Forchel, and Y. Yamamoto. Ultrafast coherent control and suppressed nuclear feedback of a single quantum dot hole qubit. *Nat. Phys.*, 7:872, 2011.
6. J. Fischer and D. Loss. Hybridization and Spin Decoherence in Heavy-Hole Quantum Dots. *Phys. Rev. Lett.*, 105:266603, 2010.
7. M.J. Biercuk, H. Uys, A. P. VanDevender, N. Shiga, W. M. Itano, and J. J. Bollinger. Optimized dynamical decoupling in a model quantum memory. *Nature*, 458:996, 2009.
8. J. Du, X. Rong, N. Zhao, Y. Wang, J. Yang, and R. B. Liu. Preserving electron spin coherence in solids by optimal dynamical decoupling. *Nature*, 461:1265, 2009.
9. H. Bluhm, S. Foletti, I. Neder, M. Rudner, D. Mahalu, V. Umansky, and A. Yacoby. Dephasing time of GaAs electron-spin qubits coupled to a nuclear bath exceeding 200 μs. *Nat. Phys.*, 7:109, 2010.
10. M. A. Nielsen and I. L. Chuang. *Quantum Computation and Quantum Information*. Cambridge University Press, 2000.
11. C. K. Hong, Z. Y. Ou, and Mandel L. Measurement of Subpicosecond Time Intervals between Two Photons by Interference. *Phys. Rev. Lett.*, 59:2044, 1987.
12. D. L. Moehring et al. Entanglement of single-atom quantum bits at a distance. *Nature*, 449:68, 2007.
13. E. Waks, K. Inoue, S. Santori, D. Fattal, J. Vuckovic, G. S. Solomon, and Y. Yamamoto. Quantum cryptography with a photon turnstile. *Nature*, 420:762, 2002.
14. H. Takesue, S. W. Nam, Q. Zhang, R. H. Hadfield, T. Honjo, K. Tamaki, and Y. Yamamoto. Quantum key distribution over a 40-dB channel loss using superconducting single-photon detectors. *Nat. Photonics*, 1:343, 2007.
15. B. B. Blinov, D. L. Moehring, L.-M. Duan, and C. Monroe. Observation of entanglement between a single trapped atom and a single photon. *Nature*, 428:153, 2004.
16. E. Knill, R. Laflamme, G. J. Milburn. A scheme for efficient quantum computation with linear optics. *Nature*, 409:46, 2001.
17. C. Santori, D. Fattal, J. Vuckovic, G. S. Solomon, and Y. Yamamoto. Indistinguishable photons from a single-photon device. *Nature*, 419:594, 2002.
18. R. B. Patel, A. J. Bennett, I. Farrer, C. A. Nicoll, D. A. Ritchie, and A. J. Shields. Two-photon interference of the emission from electrically tunable remote quantum dots. *Nat. Photonics*, 4:632, 2010.
19. H.-J. Briegel, W. Dür, J. I. Cirac, and P. Zoller. Quantum repeaters: The role of imperfect local operations in quantum communication. *Phys. Rev. Lett.*, 81:5932, 1998.
20. D. Kim, S. G. Carter, A. Greilich, A. S. Bracker, and D. Gammon. Ultrafast optical control of entanglement between two quantum-dot spins. *Nat. Phys.*, 7:223, 2011.
21. T. D. Ladd and Y. Yamamoto. Simple quantum logic gate with quantum dot cavity QED systems. *Phys. Rev. B*, 84:235307, 2011.
22. J Berezovsky, M. H. Mikkelsen, O. Gywat, N. G. Stoltz, L. A. Coldren, and D. D. Awschalom. Nondestructive Optical Measurements of a Single Electron Spin in a Quantum Dot. *Science*, 314:1916, 2006.
23. M. Atatüre, J. Dreiser, A. Badolato, and A. Imamoglu. Observation of Faraday rotation from a single confined spin. *Nat. Phys.*, 3:101, 2007.
24. A. N. Vamivakas, C.-Y. Lu, C. Matthiesen, Y. Zhao, S. Fält, A. Badolato, and M. Atatüre. Observation of spin-dependent quantum jumps via quantum dot resonance fluorescence. *Nature*, 467:297, 2010.

Appendix A
Fidelity Analysis of Coherent Control Operations

A.1 Electron Spin Control

A.1.1 Initialization and Readout Fidelity

For the data in Fig. 3.4, spin readout and initialization occurs by means of the same optical pumping laser, resonant with the $|\downarrow\rangle$-$|\uparrow\downarrow\Downarrow\rangle$ transition. This initializes the system into the $|\uparrow\rangle$-state, and allows for filtering out of a single, linearly-polarized photon from spontaneous emission along the $|\uparrow\downarrow\Downarrow\rangle$-$|\uparrow\rangle$-branch of the Λ-system as a measurement of the $|\downarrow\rangle$-population. In view of the finite single-photon extraction efficiency (approximately 3 % for our low-Q, asymmetric cavity with good backreflecting mirror), linear losses and detector efficiencies, only 0.1 % of these single photons result in a detection event (measured count rates: 80,000 counts/s for a 76 MHz repetition rate of excitation and measurement).

This optical pumping scheme can give rise to two types of errors: incorrect initialization, and incorrect spin readout. For the measurement process, ideally, the detection of a single photon should indicate that the electron spin is in the $|\downarrow\rangle$-state, with 100 % accuracy. In practice, residual leakage from the optical pumping laser due to imperfect polarization (4,000:1) and dual grating frequency filtering (1,000,000:1 at $B = 3$ T) results in approximately 2 % uncorrelated counts on the detector (measured rates: 1,000–1,500 counts per second, including detector dark counts, compared to 80,000 total counts), which is our dominant noise source. While the presence of the second Λ-system would in principle allow for optical pumping and spontaneous emission decay to occur due to absorption by the $|\uparrow\rangle$-$|\uparrow\downarrow\Uparrow\rangle$-transition, the detunings used are sufficiently large to make this process negligible compared to the direct laser leakage.

The procedure for quantifying the initialization error was first established in Ref. [1]. We apply this procedure to the time-resolved optical pumping data in Fig. 3.4, which were obtained by combining a 76 MHz train of π-rotation pulses with a constant optical pumping signal. Subtracting a constant background signal

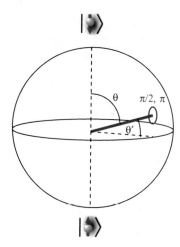

Fig. A.1 Angle conventions used in coherent control. Schematic illustration of the angle conventions used in the spin control fidelity discussion. θ represents the angle between the rotation axis (*red*) and the north pole of the Bloch sphere ($|\downarrow\rangle$-state), while θ' refers to the angle between the rotation axis and its projection on the equator of the Bloch sphere. π and $\pi/2$ refer to the coherent rotation angle, around the rotation axis (Figure reproduced with permission from [2])

corresponding to about 2 % scattered laser light, we make the assumption that the Bloch vector is optically pumped to an initial length of L_0. With the optical pumping decay time τ of 2.7 ns obtained from a least-squares fit to exponential decay (residual population: less than 1 %), we equal the minimum count rate in Fig. 3.4 with $(1-L_0)/2$. Directly after the optical π-pulse is applied, the $|\downarrow\rangle$-population is expected to increase to $(1 + L_0 D_\pi \cos 2(\theta'))/2$, where $\theta' = \pi/2 - \theta$ indicates the deviation from a rotation along an axis on the equator of the Bloch sphere, as derived above. Filling out the values for D_π and θ which are calculated below, we obtain that $L_0 = 0.92$, resulting in an initialization fidelity $F_{\text{init}} = (1 + L_0)/2 = 96\%$ for the data presented in Fig. 3.4.

A.1.2 Coherent Control Pulse Fidelity

The fidelity analysis for the ultrafast coherent spin control operations was also developed and extensively described in Ref. [1]. We use a similar axis convention as was used in that work, which is summarized in Fig. A.1.

We are particularly interested in the fidelity of the $\pi/2$ and π pulses, such as those shown in Fig. 3.9a, b. The Ramsey fringes for two $\pi/2$ pulses are shown in Fig. 3.9a, where we applied a least-squares fit of the data to a sinusoidal curve (green trace). The assumption made in order to extract the fidelity, is that the Bloch vector, initially of length L_0, shrinks by an amount $D_{\pi/2}$ after each rotation. With a scaling factor C between the $|\downarrow\rangle$-population and the obtained counts, the measured population in the $|\downarrow\rangle$-state oscillates between $CL_0(1-D_{\pi/2}^2)/2$ and $CL_0(1+D_{\pi/2}^2)/2$

after background subtraction (the optical pumping/readout laser only measures the population in the $|\downarrow\rangle$-state; the scaling factor reflects the number of photons extracted from the quantum dot in a particular amount of integration time). From the fit to the data in Fig. 3.9a, we obtain that $CL_0 = 29{,}297$ and $D_{\pi/2} = 0.893$ for the data in that experiment. The resulting $\pi/2$-pulse fidelity can therefore be estimated as $F_{\pi/2} = (1 + D_{\pi/2})/2 = 94.6\%$. In Fig. 3.9a, we see that slight asymmetry of the Ramsey fringes for delays around 80 ps or higher limits the visibility of the over-all sinusoidal least-squares fit, an effect related to dynamic nuclear polarization which was reported elsewhere [3].

We perform a similar calculation for the π-pulse fidelity. For Ramsey interference between two π-pulses (see e.g. Fig. 3.9b), one would ideally expect the fringe visibility to completely disappear. However, due to the finite duration of the rotation pulse (around 3 ps) as compared to the Larmor precession time (some 57 ps at $B = 3$ T), the π-pulses are effectively off-axis [1], with a rotation around a polar angle θ from the pole of the Bloch sphere (for an ideal, ultrafast pulse, $\theta = \pi/2$). The resulting fringes should now oscillate between $CL_0(1 - D_\pi^2)/2$ and $CL_0(1 - D_\pi^2\cos(4\theta))/2$.

Using a fit to the Ramsey fringes in Fig. 3.9b, and assuming the same values for CL_0 as obtained for $\pi/2$ pulses, we can extract that $D_\pi = 0.918$, and that $\theta = 1.451$ radians. Together, this results in a π-pulse fidelity $F_\pi = (1 - \cos(2\theta)D_\pi)/2 = 94.5\%$ for the data in that experiment.

A.2 Hole Spin Control

The fidelity analysis for ultrafast coherent hole spin control can be performed using the same methods – we refer to Ref. [4]. We focus on the fidelity of a $\pi/2$ pulse. We again assume that the Bloch vector after optical pumping starts with length L_0, and shrinks by a factor $D_{\pi/2}$ after a single $\pi/2$ pulse. The combined effect of two $\pi/2$ pulses separated by a variable Larmor precession delay τ is that the population in $|\Uparrow\rangle$ oscillates between $(1 + L_0 D_{\pi/2}^2)/2$ and $(1 - L_0 D_{\pi/2}^2)/2$ with period $2\pi/\delta_{HH}$. We use a digital lock-in procedure which automatically subtracts a background of $(1 - L_0)/2$, leading to a net oscillation between $(L_0 + L_0 D_{\pi/2}^2)/2$ and $(L_0 - L_0 D_{\pi/2}^2)/2$. With a scaling factor C connecting population in $|\Uparrow\rangle$ to the measured counts, the resulting signal oscillates between $CL_0(1 + D_{\pi/2}^2)/2$ and $CL_0(1 - D_{\pi/2}^2)/2$.

From the sinusoidal fit to the $\pi/2$ Ramsey fringes in Fig. 6.7, we obtain a value for $D_{\pi/2}$ of 0.89, which implies a fidelity $F_{\pi/2} = (1 + D_{\pi/2}) = 0.945$ as quoted in the main text. This value is comparable to currently achieved values for electron spins. The fidelity is limited by the incoherent population, which can be seen in Fig. 6.6.

A similar analysis method can be used for different pulse angles, and yields similar fidelities. For different dots, the actual value of the fidelity can vary slightly, but is consistently found to be upwards of 0.9, limited by the incoherent population induced by the optical rotation pulse.

References

1. D. Press, T. D. Ladd, B. Zhang, and Y. Yamamoto. Complete quantum control of a single quantum dot spin using ultrafast optical pulses. *Nature*, 456:218, 2008.
2. K. De Greve, L. Yu, P. L. McMahon, J. S. Pelc, C. M. Natarajan, N. Y. Kim, E. Abe, S. Maier, C. Schneider, M. Kamp, S. Höfling, R. H. Hadfield, A. Forchel, M. M. Fejer, and Y. Yamamoto. Quantum-dot spin-photon entanglement via frequency downconversion to telecom wavelength. *Nature*, 491:421, 2012.
3. T. D. Ladd, D. Press, K. De Greve, P. McMahon, B. Friess, C. Schneider, M. Kamp, S. Höfling, A. Forchel, and Y. Yamamoto. Pulsed nuclear pumping and spin diffusion in a single charged quantum dot. *Phys. Rev. Lett.*, 105:107401, 2010.
4. K. De Greve, P. L. McMahon, D. Press, T. D. Ladd, D. Bisping, C. Schneider, M. Kamp, L. Worschech, S. Höfling, A. Forchel, and Y. Yamamoto. Ultrafast coherent control and suppressed nuclear feedback of a single quantum dot hole qubit. *Nat. Phys.*, 7:872, 2011.

Appendix B
Electron Spin-Nuclear Feedback: Numerical Modelling

For an electron-charged QD, a simple and mathematically tractable nonlinear diffusion equation was derived in Ref. [1], describing the nonlinear feedback loop resulting from nuclear-dependent Larmor precession (Overhauser shift) and electron-spin dependent nuclear spin relaxation. This model can be extended to incorporate the hysteretic and asymmetric curves obtained while scanning through the optical resonance frequency of the QD (the $|\downarrow\rangle$-$|\downarrow\uparrow,\Uparrow\rangle$ transition, see Sect. B.2 and Fig. B.2).

B.1 Hysteretic and Asymmetric Electron Spin Ramsey Fringes

For a single electron spin in a QD, the strong contact hyperfine interaction makes the electron spin Larmor precession frequency very sensitive to the net polarization of the nuclear spin bath through the Overhauser shift. Conversely, in Ref. [1] it was shown that the evolution of the nuclei in the quantum dot depends on the rate of trions being generated, as the unpaired hole in the electron spin trion state allows for quasi energy-conserving nuclear spin flips to occur. This effect, together with background nuclear spin diffusion, can be modelled as a nonlinear diffusion equation for the average Overhauser shift ω:

$$\frac{\partial \omega}{\partial t} = -\kappa \omega + \alpha \frac{\partial C(\omega, \tau)}{\partial \omega}, \tag{B.1}$$

where $C(\omega, \tau)$ is the trion generation rate in the experiment. For Ramsey fringes, $C(\omega, \tau)$ can be calculated by analyzing the pulse pattern (see Fig. B.1b, inset). In particular, a CW optical pumping pulse is interspersed with two $\pi/2$ rotation pulses, separated by a variable delay τ. Denoting the average spin polarization by S, the spin-up probability by $P(\uparrow)$ (equal to $(1+S)/2$) and the spin-down probability by

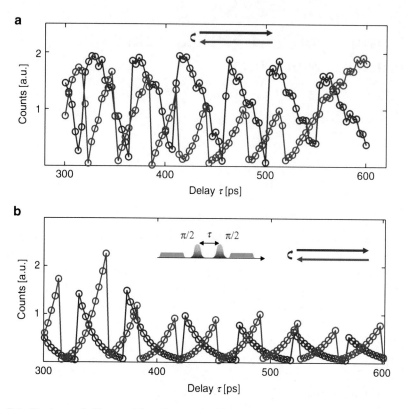

Fig. B.1 Electron spin Ramsey fringe hysteresis. (**a**) Asymmetric and hysteretic Ramsey fringes for an electron spin (measured; from Ref. [1]). The *green* and *blue* arrows indicate the scanning direction of the pulse delay τ. (**b**) Asymmetric and hysteretic electron spin Ramsey fringes as predicted by the model in Ref. [1]. *Inset*: scan direction and pulse timing (Figures reproduced from [2])

$P(\downarrow)$, we can relate the spin polarization S_{after} right after optical pumping for a time T to the polarization S_{before} before the arrival of the optical pumping pulse:

$$S_{\text{after}} = \sum_{m=\uparrow,\downarrow} \left[\frac{1}{2} P(\uparrow|m) P_{\text{before}}(m) - \frac{1}{2} P(\downarrow|m) P_{\text{before}}(m) \right]$$
$$= \frac{(1 - e^{-\beta(\omega)T})}{2} + S_{\text{before}} e^{-\beta(\omega)T}. \quad (B.2)$$

Here, $\beta(\omega)$ represents the optical pumping/absorption profile, assumed to be Gaussian. In addition, the interference of the two $\pi/2$ rotation pulses separated by a delay τ yields

$$S_{\text{before}} = -\cos((\omega_0 + \omega)\tau) S_{\text{after}}, \quad (B.3)$$

where ω_0 represents the Larmor precession in the absence of any Overhauser effects. Together, this results in a trion generation rate $C(\omega, \tau)$:

$$C(\omega, \tau) = S_{\text{after}} - S_{\text{before}}$$
$$= \frac{1}{2} \frac{(1 - e^{-\beta(\omega)T})\{1 + \cos[(\omega_0 + \omega)\tau]\}}{1 + \cos[(\omega_0 + \omega)\tau]e^{-\beta(\omega)T}}. \quad \text{(B.4)}$$

Equation (B.1) is derived by invoking a separation of timescales [1], and can be solved to yield steady-state solutions $\omega_f(\tau)$. The integration times used in the experiment (few seconds for each different value of τ) were found to be sufficient for reaching these steady-state solutions. Due to the nonlinearity, the solutions depend on the initial conditions, and therefore on the direction in which τ is varied. Figure B.1a, b compare the experimentally obtained Ramsey fringes with those obtained by numerically solving Eq. (B.1). In Fig. B.1b, $\beta(\omega)$ was assumed to be Gaussian ($\beta_0 e^{-(\omega^2/2\sigma^2)}$) with $\sigma/2\pi = 1.6\,\text{GHz}$), $\kappa = 10\,\text{s}^{-1}$ and $\kappa/\alpha = 10^4\,\text{ps}^2$.

B.2 Hysteretic and Asymmetric Electron Spin CW Resonance Scanning

In Refs. [3, 4] hysteretic effects were observed while scanning a narrowband CW laser through the QD optical resonance frequency. In Ref. [4], a single electron-charged QD was studied using Coherent Population Trapping in Voigt geometry. In order to avoid optical pumping into one of the electron spin ground states, one laser was kept fixed, while another laser was scanned through the QD resonance wavelength. Trion-induced nuclear spin flips lead to a dragging of the resonance wavelength upon scanning the laser frequency, and the consequent hysteresis.

A similar effect was observed for a single electron-charged QD in our system. Instead of having two narrowband lasers, a π rotation pulse was used to compensate for optical pumping, and a narrowband CW-laser (few MHz linewidth) was scanned through the QD resonance wavelength ($|\downarrow\rangle$-$|\downarrow\uparrow, \Uparrow\rangle$-transition [1]). Figure B.2a illustrates the hysteresis and asymmetry upon scanning the CW-laser in different directions.

Equation (B.1) can again be used to model the Overhauser shift ω. The difference with the Ramsey fringe hysteresis lies in the trion generation rate $C(\omega, \omega_{\text{las}}, \theta)$, where ω_{las} stands for the laser frequency, and θ for the rotation angle of the single pulse used in the experiment (π in our case). The pulse sequence used is shown in Fig. B.2b, inset. After optical pumping for a time T, we still have that

$$S_{\text{after}} = \frac{(1 - e^{-\beta(\omega, \omega_{\text{las}}, \omega_{\text{res}})T})}{2} + S_{\text{before}} e^{-\beta(\omega, \omega_{\text{las}}, \omega_{\text{res}})T}, \quad \text{(B.5)}$$

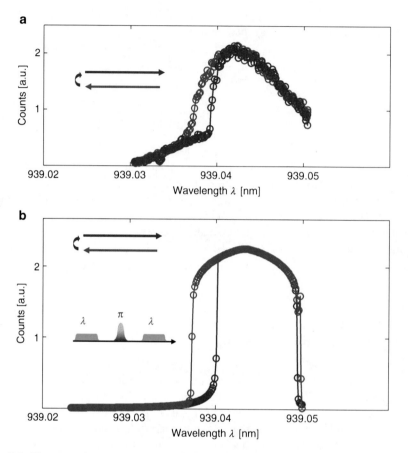

Fig. B.2 Electron spin resonant scanning hysteresis. (a) Asymmetric and hysteretic resonance scan for an electron spin (experimental). The *green* and *blue* circles indicate the wavelength scanning direction. (b) Asymmetric and hysteretic electron spin resonance scan, as predicted by an extension of the model in Ref. [1]. *Inset*: scan direction and pulse timing (Figures reproduced from [2])

where ω_{las} and ω_{res} respectively stand for the laser frequency and the QD resonance frequency in the absence of nuclear spin effects. However, the single rotation pulse with angle θ has now a different effect on the spin polarization:

$$S_{\text{before}} = \cos(\theta) S_{\text{after}}. \tag{B.6}$$

This results in a net trion generation rate $C(\omega, \omega_{\text{las}}, \theta)$:

$$\begin{aligned} C(\omega, \omega_{\text{las}}, \theta) &= S_{\text{after}} - S_{\text{before}} \\ &= \frac{1}{2} \frac{(1 - e^{-\beta(\omega, \omega_{\text{las}}, \omega_{\text{res}})T})[1 - \cos(\theta)]}{1 - \cos(\theta) e^{-\beta(\omega, \omega_{\text{las}}, \omega_{\text{res}})T}}. \end{aligned} \tag{B.7}$$

We can again obtain steady-state values ω_f from Eq. (B.1). Whether or not steady state is obtained, however, depends critically on the scan speed – a dependence we also notice experimentally. We assume a Lorentzian QD linewidth ($\beta(\omega) = \beta_0/(1 + (\omega_{las} - \omega_{res} - \omega)^2/\sigma^2)$, $\sigma/2\pi = 200$ MHz). Other lineshapes (Gaussian, Voigt) yield qualitatively similar results. While the exact resulting lineshape depends critically on the initial conditions and details of the QD and experiment (initial Overhauser shift ω_0, scan speed, lineshape, etc.), the qualitative features are well reproduced in this model; Fig. B.2b shows the numerical solution to Eq. (B.1). κ is estimated as $8.5\,\mathrm{s}^{-1}$, and $\kappa/\alpha = 2.8 \times 10^4\,\mathrm{ps}^2$.

B.3 Nuclear Feedback: Comparison Between Electron and Hole

For the hysteretic effects of a single electron spin coupled to the nuclear spins in the QD, the average Overhauser contribution to the Larmor precession frequency can be extracted from the model described above. Figure B.3 compares the different behavior of electrons and holes; Fig. B.3a displays the Overhauser shift extracted

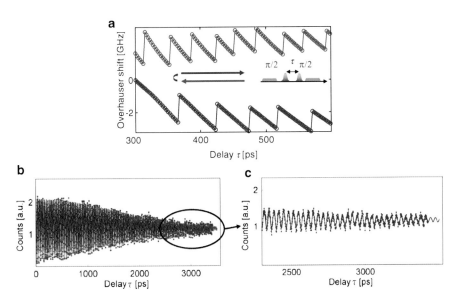

Fig. B.3 **Electron and hole spin Overhauser shifts compared.** (a) Modeled average Overhauser shift for hysteretic Ramsey fringes of a single electron spin; the *green* and *blue* circles indicate the wavelength scanning direction, as indicated by the *arrows*. (b) Time-averaged dephasing of a single hole spin; *blue*: raw data, *red*: fit to a sinusoid with Gaussian envelope. No variance of the average Overhauser shift was observed. (c) Zoomed-in version of (b) – note that the phase of the fringes remains constant over the entire scan range (Figures reproduced from [2])

from the model. As our model for the electron spin captures only the average Overhauser shift [1,5], one conservative way of estimating the error on the obtained values is to assume that there is no additional narrowing of the nuclear spin distribution due to the development of nuclear spin polarization [1]. In that case, the T_2^* value of 1.71 ns obtained in Ref. [6] can be used to estimate the variance on the Overhauser shift due to time-ensemble effects, yielding $\sigma_\omega/2\pi = \sqrt{2}/(2\pi T_2^*) = 130$ MHz. We can therefore estimate the maximum Overhauser shifts for a single electron spin due to the interaction with the nuclei at 3 ± 0.13 GHz. The resonance scan model predicts a similar, or slightly reduced, maximum Overhauser shift.

A single hole spin does not display any hysteresis or nonlinearity in either the Ramsey fringe or resonance scanning experiments. Moreover, compared to the indirect method of extracting Overhauser shifts through modelling based on Eq. (B.1), a more accurate estimate of the Overhauser shift can be obtained from the phase of the Ramsey fringes. That phase equals $(\omega_0 + \omega)\tau$, and the Ramsey fringes are shown in Fig. B.3b, c, together with a sinusoidal fit with Gaussian envelope (red curve). The raw data hardly deviate from the fit, except for very long delays, where noise effects dominate. Even with the noise, the deviation is at most 0.5–1 radians for a total delay τ of 3.5 ns, leading to a maximum Overhauser shift $\omega/2\pi$ of 40 ± 100 MHz. We may bound any possible hole Overhauser shifts by supposing they are masked by experimental noise. Here, the width of the time-averaged Larmor precession frequency distribution leading to T_2^*-decay results in an uncertainty $\sigma_\omega/2\pi = \sqrt{2}/(2\pi T_2^*) = 100$ MHz, using our experimentally observed T_2^* value of 2.3 ns. We emphasize that this is a worst-case estimate, for the case in which nuclear effects would limit the time-averaged dephasing, which we consider unlikely.

Comparing these two values, we see that the developed Overhauser shift for the hole spin is at least 30 times smaller than that for the electron spin, where the factor of 30 is limited by experimental noise and T_2^* effects of the hole spin. While the Overhauser shift depends both on the developed nuclear spin polarization and the sensitivity of the hole spin to that nuclear spin polarization, this significant reduction of the measured Overhauser shift illustrates the suppression of feedback effects between the nuclear spin bath and the hole spin.

References

1. T. D. Ladd, D. Press, K. De Greve, P. McMahon, B. Friess, C. Schneider, M. Kamp, S. Höfling, A. Forchel, and Y. Yamamoto. Pulsed nuclear pumping and spin diffusion in a single charged quantum dot. *Phys. Rev. Lett.*, 105:107401, 2010.
2. K. De Greve, P. L. McMahon, D. Press, T. D. Ladd, D. Bisping, C. Schneider, M. Kamp, L. Worschech, S. Höfling, A. Forchel, and Y. Yamamoto. Ultrafast coherent control and suppressed nuclear feedback of a single quantum dot hole qubit. *Nat. Phys.*, 7:872, 2011.
3. C. Latta *et al*. Confluence of resonant laser excitation and bidirectional quantum-dot nuclear-spin polarization. *Nat. Phys.*, 5:758, 2009.

References

4. X. Xu et al. Optically controlled locking of the nuclear field via coherent dark-state spectroscopy. *Nature*, 459(4):1105, 2009.
5. T. D. Ladd, D. Press, K. De Greve, P. McMahon, B. Friess, C. Schneider, M. Kamp, S. Höfling, A. Forchel, and Y. Yamamoto. Nuclear feedback in a single quantum dot under pulsed optical control. arXiv:1008.0912v1.
6. D. Press, K. De Greve, P. McMahon, T. D. Ladd, B. Friess, C. Schneider, M. Kamp, S. Höfling, A. Forchel, and Y. Yamamoto. Ultrafast optical spin echo in a single quantum dot. *Nat. Photonics*, 4:367, 2010.

Appendix C
Extraction of Heavy-Light Hole Mixing Through Photoluminescence

The heavy-hole light-hole mixing can be quantified using photoluminescence (PL), as reported in Ref. [2]. We performed a similar analysis for the hole-charged quantum dot studies presented in Chap. 6. Without any external magnetic field, the polarization of the emitted PL contains information about the hole spin eigenstates. In particular, strain and quantum dot asymmetry result in a small amount of HH-LH mixing. The resulting hole spin ground states along the growth direction (z), $|\Uparrow\rangle$ and $|\Downarrow\rangle$ can be modeled as:

$$|\Uparrow\rangle = [|\Psi_{+3/2}\rangle + \eta^+ |\Psi_{-1/2}\rangle] / \sqrt{1 + |\eta|^2}$$

$$|\Downarrow\rangle = [|\Psi_{-3/2}\rangle + \eta^- |\Psi_{+1/2}\rangle] / \sqrt{1 + |\eta|^2}. \quad (C.1)$$

Here, $|\Psi_{\pm 3/2}\rangle$ and $|\Psi_{\pm 1/2}\rangle$ represent the HH and LH states respectively, and $\eta^\pm = |\eta| e^{\pm i\xi}$ (ξ indicates an orientation of high symmetry, e.g. determined by a preferential strain direction – see Ref. [2] for further details). Using now a similar analysis as in Sect. 2.3, LH inmixing reflects itself in a slightly elliptical polarization of the hole-charged PL. Setting $\xi = 0$ (this assumes symmetry along the x-direction; the extension for $\xi \neq 0$ is straightforward, and would result in axes for the resulting elliptical polarization that are *not* along the x- or y-direction), we have the following polarization for decay from the hole trion states to the hole ground states:

$$|\Uparrow\Downarrow\uparrow\rangle \rightarrow |\Uparrow\rangle : \sigma^- + \frac{|\eta|}{\sqrt{3}} \sigma^+ \quad (C.2)$$

$$|\Uparrow\Downarrow\downarrow\rangle \rightarrow |\Downarrow\rangle : \sigma^+ + \frac{|\eta|}{\sqrt{3}} \sigma^- \quad (C.3)$$

This elliptical polarization can be visualized in a polarization-resolved photoluminescence experiment: for a statistical mixture of both decay processes, the collected light along particular polarizations will no longer be constant. We refer to Fig. C.1,

C Extraction of Heavy-Light Hole Mixing Through Photoluminescence

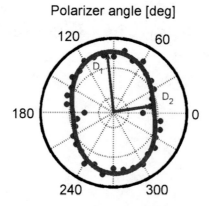

Fig. C.1 Polarization angle dependence of the emitted photoluminescence (PL) of a hole-charged QD at 0 magnetic field. *Blue dots*: raw data; *red curve*: least squares fit of the elliptical polarization. The distance from the origin indicates the relative intensity of the emitted PL for a particular polarization angle. D_1 and D_2 are the main axes of the resulting elliptical polarization dependence (see text). Note that the data were taken for polarization angles between 0° and 180°, and copied for the 180–360° trajectory in view of the inversion symmetry of the system. The discontinuity at 0° is a systematic experimental artifact (Figure reproduced from [1])

which reflects the ellipticity of the emitted light for the hole-charged quantum dots analyzed in Chap. 6. From the ratio, R, between the two axes of the elliptical polarization ($D_{1,2}$), we obtain a measure for the amount of inmixing:

$$R = (\sqrt{3} - |\eta|)^2 / (\sqrt{3} + |\eta|)^2. \tag{C.4}$$

For the data in Fig. C.1, the inmixing could be estimated at $\eta \sim 17\%$.

References

1. K. De Greve, P. L. McMahon, D. Press, T. D. Ladd, D. Bisping, C. Schneider, M. Kamp, L. Worschech, S. Höfling, A. Forchel, and Y. Yamamoto. Ultrafast coherent control and suppressed nuclear feedback of a single quantum dot hole qubit. *Nat. Phys.*, 7:872, 2011.
2. T. Belhadj et al. Impact of heavy hole-light hole coupling on optical selection rules in GaAs. *Appl. Phys. Lett.*, 97:051111, 2010.

Appendix D
Numerical Modeling of Ultrafast Coherent Hole Rotations

The Rabi-oscillations for a single hole qubit presented in Fig. 6.6 can be modeled using the AC-Stark shift model developed in Sect. 3.1.2. As the pulse duration of 3.67 ps is much shorter than the Larmor precession frequency $\delta_{HH}/2\pi = 30.2$ GHz, one can look at the interaction in the basis of the light pulse (z-basis as indicated in Fig. D.1a). In this basis, the magnetic field results in an off-diagonal coupling between the hole spins, indicated by B_x in Fig. D.1b. However, given that the pulse is much faster than the Larmor-precession, the z-basis spins can be considered as effectively degenerate, and the magnetic field can be approximately neglected in the remainder of the analysis.

For perfect selection rules and ideally circularly polarized light pulses, only one of the z-basis hole spin states is coupled to the trion states; the other state is dark. For realistic quantum dots, imperfect selection rules and limited control over the exact polarization of the light pulse inside the cavity lead to both hole spin ground states being coupled to the trion states. The coupling strengths $\Omega_{1,2}$ are indicated in Fig. D.1b; even for realistic quantum dots with non-negligible amounts of heavy- and light-hole mixing, one coupling strength is typically much larger than the other. For a detuning Δ (340 GHz in our case) larger than the pulse bandwidth, the pulse mixes the hole spin ground state and its excited trion state. The effect of the time-dependent mixing is a time-dependent AC-Stark-shift $\delta_{1,2}(t)$, given by:

$$\delta_{1,2}(t) = \frac{1}{2}\sqrt{\Delta^2 + |\Omega_{1,2}(t)|^2} - \frac{\Delta}{2}. \tag{D.1}$$

A hole spin initialized in the x-basis due to the magnetic field can be written as a superposition of z-basis states with equal weight. The effect of the pulse is then to AC-Stark-shift these states by a different amount, leading to rotation pulse power (P_{rot}) dependent Rabi oscillations with net rotation angle:

$$\theta = \int dt \left[\delta_1(t) - \delta_2(t)\right] = \frac{1}{2}\int dt \left[\sqrt{\Delta^2 + |\Omega_1(t)|^2} - \sqrt{\Delta^2 + |\Omega_2(t)|^2}\right]. \tag{D.2}$$

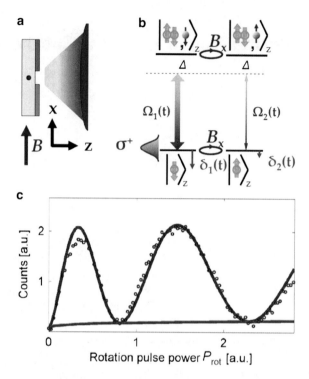

Fig. D.1 AC-Stark model for hole qubit rotations. (**a**) Geometry and axis convention used in the experiment. The magnetic field is oriented along x, while the laser pulse is aligned to the growth direction z. (**b**) AC-Stark shift in the Z-basis: $|\Downarrow\rangle$ and $|\Uparrow\rangle$ are the hole spin ground states, while $|\Downarrow\Uparrow,\downarrow\rangle$ and $|\Downarrow\Uparrow,\uparrow\rangle$ represent the trion states. Δ represents the detuning, and the circularly polarized laser pulse couples the ground states to the excited states ($\Omega_{1,2}(t)$), resulting in AC-Stark shifts $\delta_{1,2}(t)$. (**c**) Rabi oscillations fit through the AC-Stark model. *Blue* circles raw data; *red*: AC-Stark shift predicted Rabi oscillations, on top of an incoherent background (*green*) (Figures reproduced from [1])

Here, the integration is over the duration of a single rotation pulse. Figure D.1c illustrates the predicted Rabi oscillations in this AC-Stark framework. The data show an incoherent background ($\sim P_{\text{rot}}^{0.65}$) which is shown as the green curve in this figure. After subtracting the background, a least-squares fit extracted the amplitude of the Rabi oscillations. The pulse shape was modelled as Gaussian, with a measured FWHM of 3.67 ps, and for the detuning, the measured value of 340 GHz was used. The best fit was obtained for a ratio $|\Omega_1(t)|^2/|\Omega_2(t)|^2 = 3.7$, and is indicated by the red curve. The model fits the data very well, with the exception of the height of the first peak. This deviation can be attributed to the still finite duration of the laser pulse, and our neglecting the Larmor precession in this model. In Ref. [2] we demonstrated how the combined effect of Larmor precession and pulse-induced Rabi oscillations leads to an effective rotation axis that is in between the laser pulse (z-axis) and the magnetic field axis (x), leading to a reduced height of the first π

pulse. A full time-dependent coherent simulation can qualitatively reproduce the reduced height. The background and upward trend, however, cannot be reproduced by this simulation. Its origin is currently unknown, although it might be related to a change in the optimum bias position of the QD for high rotation pulse powers as reported above.

References

1. K. De Greve, P. L. McMahon, D. Press, T. D. Ladd, D. Bisping, C. Schneider, M. Kamp, L. Worschech, S. Höfling, A. Forchel, and Y. Yamamoto. Ultrafast coherent control and suppressed nuclear feedback of a single quantum dot hole qubit. *Nat. Phys.*, 7:872, 2011.
2. D. Press, T. D. Ladd, B. Zhang, and Y. Yamamoto. Complete quantum control of a single quantum dot spin using ultrafast optical pulses. *Nature*, 456:218, 2008.

Appendix E
Hole Spin Device Design

For the hole-spin studies reported in Chap. 6, two different types of samples were studied: δ-doped samples, and charge-tuneable devices. The δ-doped samples contain about $1.5 \times 10^{10}\,\text{cm}^{-2}$ self-assembled quantum dots, and the charge-tuneable samples about $7 \times 10^9\,\text{cm}^{-2}$. For both types of samples, the quantum dots (QDs) were grown using the Stranski-Krastanov method. The Indium flushing and partial capping technique used during the QD growth [2] leads to the formation of flattened QDs, with an approximate height of 2 nm, and a base length of $\sim$25 nm. The detailed layer structures are provided in Fig. E.1. For both types of samples, the QDs are embedded in a planar microcavity, consisting of Distributed Bragg Reflector (DBR) mirrors. The top and bottom mirrors consist of 5 and 25 pairs of AlAs/GaAs $\lambda/4$ layers respectively. The resulting quality factor is around 200, and helps both in increasing the signal strength (directing the emitted light upward) and reducing the noise (enabling the use of lower laser power, and therefore reducing the noise due to scattered laser light).

For the δ-doped samples, a carbon δ-doping layer is used, located 10 nm below the QDs. The δ-doping concentration is approximately $1.2 \times 10^{11}\,\text{cm}^{-2}$, and leads to a fraction of the QDs being charged with a single hole; we perform magneto-PL measurements in order to identify those QDs that are charged. For the charge-tuneable samples, deterministic charging occurs by embedding the QDs into a p-i-n-diode structure. The bottom DBR, as well as part of the cavity, is p-doped ($\geq 10^{18}\,\text{cm}^{-3}$), while the top DBR is n-doped ($\geq 10^{18}\,\text{cm}^{-3}$). The i-layer consists of two parts: a 25 nm i-GaAs layer acting as a tunnel barrier between the QDs and the p-layer [3, 4], and a 120 nm layer separating the QDs from the n-contact. Inside the latter, we incorporated a 110 nm i-AlAs/GaAs superlattice (20 layers) to prevent charging from the n-layer [4]. A back contact (not shown) allows for biasing of the substrate, while a metal shadow mask also serves as a contact to the n-doped layer. Apertures in the metal mask provide optical access to the QDs, at the expense of a reduced net bias over the QDs: given the relatively large width of the metal mask apertures (16 μm), the exact bias over a QD depends on the lateral position of that QD within the aperture shadow. In particular, for QDs near the center of the

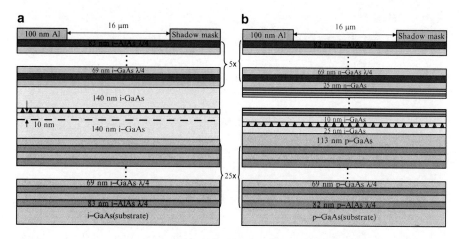

Fig. E.1 Detailed layer structure of the hole devices used. (**a**) Detailed layer structure of the δ-doped samples. The δ-doping layer (*dashed line*) is located 10 nm below the quantum dots. (**b**) Detailed layer structure of the charge-tuneable devices used in the hole spin experiment. Two DBR layer stacks form an asymmetric cavity, in which a p-i-n-diode is embedded. QDs (*brown triangles*) are in tunnel contact (25 nm i-GaAs) with a hole reservoir. The bias voltage is applied over the Al shadow mask, and a bottom contact (not shown) (Figures reproduced from [1])

aperture, part of the applied bias voltage will result in a resistive voltage drop inside the n-layer, reducing the net bias over the QD. In addition, Schottky-barrier effects at the metal mask-DBR interface further reduce the effective QD bias for a given applied bias voltage.

We can calculate the band structure and energy levels of the QD in a full three-dimensional simulation, using the 3D simulation tool "nextnano" [5]. The results are shown in Fig. E.2a, b. We assume an InAs QD with an approximate height of 2 nm, and a base length of 25 nm, and account for a finite amount of In-Ga intermixing. The resulting QD emission wavelength is around 940 nm. The two most tightly bound states in both the conduction band (electron charging) and valence band (hole charging) are indicated by the red and green dashed lines. We calculate both the HH and LH subbands, though the HH band is by far the most important. The two most tightly bound HH states in the QD are separated by ~ 14 meV, and are located 199 meV above the GaAs valence band.

Next, the band structure of the charge-tuneable devices can be calculated – see Fig. E.2c, d. Without externally applied bias, the built-in diode voltage leads to a band bending (black curve), where the hole Fermi-level is located ~ 230 meV above the GaAs valence band right at the position of the QD. In order to keep our calculations tractable, we split the problem into two subproblems: a full 3D calculation of the band line-up of the QD HH bound states, and a 1D calculation of the band bending of the entire device. Hence, without applied bias, the HH QD states are located below the Fermi-level, making hole charging of the QD energetically unfavorable. At some positive bias, the offset between the Fermi-level and the HH

E Hole Spin Device Design

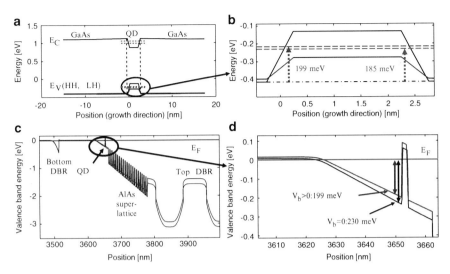

Fig. E.2 Calculation of the band line-up in the charge-tuneable hole devices. (**a**) Calculated energy structure (3D) of the InAs QD. $E_{C,V}$ are the conduction and valence bands respectively (the latter is calculated for the heavy (HH, *black*) and light hole (LH, *blue*) subbands). The two most tightly bound energy states in the conduction and valence bands are indicated by the *red* and *green dashed lines*. (**b**) Zoomed-in version of (**a**). A flattened QD was assumed, with 2 nm height, 25 nm base length, and up to 50 % In-Ga intermixing. The most tightly bound HH state is located 199 meV above the GaAs valence band, and the second-highest state is split off by ~14 meV. (**c**) Band line-up for zero (*black*) and positive (*blue*) bias (note that only the p-i-n-diode region is shown, not the entire DBR microcavity). E_F: Fermi-level. (**d**) Zoomed-in version of (**c**): for zero applied bias, the GaAs valence band lies 230 meV below the Fermi-level, and the most tightly bound HH state therefore lies below the Fermi-level. For a positive bias, the offset between the Fermi-level and the GaAs valence band is reduced; when this offset equals the 199 meV separation with the most tightly bound HH state, resonant tunneling can result in hole-charging of the QD (Figures reproduced from [1])

QD state vanishes (blue curve), with resonant tunneling allowing for deterministic charging of the QD. Numerically, this bias voltage is around 200 mV, though we emphasize that this is the real bias over the p-i-n-diode near the QD, which is often less than the applied voltage between the device contacts, especially for QDs located near the center of the mask aperture (DC Stark shifts and Schottky-barrier effects account for an additional offset). Pauli- and Coulomb-blockade effects subsequently lead to a stable voltage plateau where single-hole charging is possible.

In view of the relatively large QD density, about 50 QDs are located within our diffraction-limited laser spot (charge-tuneable devices; for the δ-doped samples, some 150 QDs), and several of those are resonant with the microcavity. In Fig. 6.4a we show the photoluminescence (PL) as a function of applied bias voltage for a particular QD from a charge-tuneable device (above-band excitation, $\lambda = 785$ nm). A magnetic field of 6 T in Voigt geometry splits the transitions, and identification of the respective lines is made easier through the particular fine structure of transitions from charged and uncharged QDs [6]. The charged QDs display a fourfold split of the PL for large magnetic field – see Fig. 6.4b, where the magneto-PL of a hole-charged QD is shown (note that two of the four lines overlap due to limited spectrometer resolution, which can be seen as an apparent increase in brightness of the center line); it is this same particular signature that also allows us to identify the charged QDs in the δ-doped samples. The dependence on the pumping power allows us to separate excitonic emission lines from lines due to multi-excitonic complexes. The inhomogeneity in size and composition of the different QDs, together with the expected spectral line-up of the different charge states of a single QD, allow us to identify the lines in Fig. 6.4a. As expected, increasing the QD bias leads to a transition from an uncharged to a charged state. However, and as reported in Ref. [3], we see a significant overlap between the respective voltage plateaus of the charged and uncharged QD state, which can be attributed to the relatively slow tunneling of the hole in our QDs. In addition, we notice that the exact position of the voltage plateaus depends on the amount of optical power used. Both above-band and resonant CW-excitation (as well as below-band modelocked (ML) laser pulses used for coherent spin rotations) can alter the bias voltage by as much as 0.1–0.2 V – we attribute this to residual absorption in the vicinity of the QD, which leads to the generation of charged carriers that can shift the QD energy ("DC"-Stark shift).

References

1. K. De Greve, P. L. McMahon, D. Press, T. D. Ladd, D. Bisping, C. Schneider, M. Kamp, L. Worschech, S. Höfling, A. Forchel, and Y. Yamamoto. Ultrafast coherent control and suppressed nuclear feedback of a single quantum dot hole qubit. *Nat. Phys.*, 7:872, 2011.
2. J. M. García, T. Mankad, P. O. Holtz, P. J. Wellman, and P. M. Petroff. Electronic states tuning of InAs self-assembled quantum dots. *Appl. Phys. Lett.*, 72:3172–3174, 1998.
3. B. D. Gerardot *et al*. Optical pumping of a single hole spin in a quantum dot. *Nature*, 451:441, 2008.
4. D. Brunner, B. D. Gerardot, P. A. Dalgarno, G. Wüst, K. Karrai, N. G. Stoltz, P. M. Petroff, and R. J. Warburton. A coherent single-hole spin in a semiconductor. *Science*, 325(5936):70–72, 2009.
5. http://www.nextnano.de/.
6. M. Bayer *et al*. Fine structure of neutral and charged excitons in self-assembled In(Ga)As/(Al)GaAs quantum dots. *Phys. Rev. B*, 65:195315, 2002.

Appendix F
Ultrafast Quantum Eraser: Expected Visibility/Fidelity

In order to quantify the effectiveness of the ultrafast downconversion technique for time-resolved detection of $\sigma^{+,-}$-downconverted photons in Chap. 7, its timing resolution needs to be compared to the Zeeman energy corresponding to $\delta\omega = 2\pi \times 17.6\,\text{GHz}$ at 3 T (57 ps Larmor precession period). The timing resolution depends on the duration of the 2.2 μm pump pulse, which is shown to be 8 ps or better (the detailed pulse shape is a complicated function of the amount of power and the pulse shapes used in order to generate the 2.2 μm pulse), and is typically between 5 and 8 ps; we will approximate it as Gaussian for the remainder of the discussion, in reasonable agreement with the cross-correlation data in Fig. 7.3. The net effect of this finite timing resolution is a statistical mixture of ideally, infinitely-time-resolved $\sigma^{+,-}$-downconverted photons, and the overlap (fidelity) of the statistical mixture with the infinitely time-resolved case can be computed. This effect is illustrated in Fig. F.1a, and the effect on the visibility of the observed Ramsey fringes is illustrated in Fig. F.1b. For a sinusoidal fringe of period 57 ps, convolution with a Gaussian timing response function with time constant of 8 ps leads to a fringe visibility of approximately 90.7 %, yielding an upper bound of the potential visibility of the correlations in the rotated basis. Different magnetic fields and consequent Zeeman splittings result in different potential visibilities for the same absolute timing resolution. For a 6 T field, for example, with $\delta\omega = 2\pi \times 35.2\,\text{GHz}$, the expected visibility would be limited to 68 % using a similar analysis. While lower fields result in better time-filtering, the practical limit on the magnetic field is given by the difficulty of measuring the spin state in our system at fields below about 2.5 T. In addition, dynamic nuclear polarization [2] and T_2^*-effects [3] can restrict the visibility of the Ramsey fringes. However, as was shown before, all-optical spin echo techniques can be used to overcome these effects [3].

In practice, the visibility of the spin-photon correlations is limited by residual noise from the downconversion process (leaked/converted 2.2 μm pump light as well as SNSPD dark counts) and imperfectly filtered, reflected excitation laser light. The exact amounts vary slightly from experimental run to experimental run, but result consistently in overall signal-to-noise ratios between 4:1 (worst case, for rotated

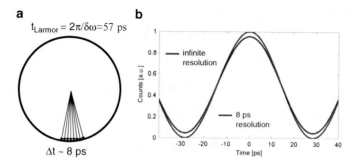

Fig. F.1 Timing resolution of the downconversion setup. (**a**) Schematic illustration of the effect of imperfect timing resolution. Instead of having an infinitely accurate time reference for the start of the Larmor precession, timing uncertainty gives rise to a statistical mixture of possible precession start-times, leading to an uncertainty cone in the Bloch sphere. This uncertainty cone needs to be compared to the precession time. (**b**) When mapped into a Ramsey fringe by applying a $\pi/2$ pulse, the timing uncertainty results in an inherent loss of visibility of the fringes. *Blue curve*: infinite timing accuracy; *red curve*: 8 ps accuracy, vs. 57 ps Larmor precession period. The resulting, theoretically maximal visibility is approximately 90.7 % (Reproduced from [1])

basis measurements without cross-polarization of the excitation pulse) and 10:1 (best case, for computational basis measurements where good cross-polarization reduced the effects of reflected excitation pulses). These signal to noise ratios limit the practical visibility of the rotated basis correlations to some 80 %.

References

1. K. De Greve, L. Yu, P. L. McMahon, J. S. Pelc, C. M. Natarajan, N. Y. Kim, E. Abe, S. Maier, C. Schneider, M. Kamp, S. Höfling, R. H. Hadfield, A. Forchel, M. M. Fejer, and Y. Yamamoto. Quantum-dot spin-photon entanglement via frequency downconversion to telecom wavelength. *Nature*, 491:421, 2012.
2. T. D. Ladd, D. Press, K. De Greve, P. McMahon, B. Friess, C. Schneider, M. Kamp, S. Höfling, A. Forchel, and Y. Yamamoto. Pulsed nuclear pumping and spin diffusion in a single charged quantum dot. *Phys. Rev. Lett.*, 105:107401, 2010.
3. D. Press, K. De Greve, P. McMahon, T. D. Ladd, B. Friess, C. Schneider, M. Kamp, S. Höfling, A. Forchel, and Y. Yamamoto. Ultrafast optical spin echo in a single quantum dot. *Nat. Photonics*, 4:367, 2010.

Printed by Publishers' Graphics LLC
MLSI130628.15.16.44